BLASTOFF!

BLASTOFF!

JUPITER

by Martin Schwabacher

BENCHMARK BOOKS

MARSHALL CAVENDISH

NEW YORK

With special thanks to Roy A. Gallant, Southworth Planetarium, University of Southern Maine, for his careful review of the manuscript.

Benchmark Books
Marshall Cavendish Corporation
99 White Plains Road
Tarrytown, NY 10591-9001
www.marshallcavendish.com

Library of Congress Cataloging-in-Publication Data
Schwabacher, Martin.
Jupiter / by Martin Schwabacher.
p. cm. — (Blastoff!)
Includes bibliographical references and index.
ISBN 0-7614-1236-0
1. Jupiter (Planet)—Juvenile literature. [1. Jupiter (Planet)] I. Title. II. Series.
QB661 .S38 2001   523.45—dc21   2001 025640

Printed in Italy
1 3 5 6 4 2

Photo Research by Anne Burns Images
Cover Photo: NASA

The photographs in this book are used by permission and through the courtesy of: Photo Researchers: NASA, 7, 19, 33, 35, 43, 45, 47(top & bottom), 49; Science Photo Library, 8, 39; Space Telescope Science Institute/NASA/Science Photo Library, 10, 54; David Hardy/Science Photo Library, 11; Lynette Cook/Science Photo Library, 13; Roger Harris/Science Photo Library, 16; U.S. Geological Survey/NASA/Science Source, 26; Jack Finch, 36; Mark Garlick, 51; Julian Baum, 55; T. Stevens & P. McKinley/Pacific Northwest Laboratory, 56. Jet Propulsion Lab: 9, 28, 29, 31, 37, 42, 44. Corbis: 21, 24; Diego Lezama Orezzoli, 17; Paul Almasy, 20. NASA: 23, 32. Photri: 40, 50, 53.

Book design by Clair Moritz-Magnesio

CONTENTS

1

ALMOST A STAR

Jupiter is the largest planet in the Solar System. In fact, it is bigger than all the other planets combined. In many ways, Jupiter resembles the Sun—the star around which Earth and the other planets orbit. With its twenty-eight moons, Jupiter is like a mini–Solar System all its own.

It is not only Jupiter's enormous size that makes it resemble a star. Jupiter is a big ball of gas made of more than 98 percent hydrogen and helium—just like the Sun. If Jupiter were eighty times bigger, it would *be* a star. If it were thirteen times bigger, it would be considered a brown dwarf, a baby star that failed to grow big enough to start glowing.

To get an idea of how big Jupiter actually is, consider its four largest moons. They would be good-sized planets in their own right, if they circled the Sun instead of Jupiter. Ganymede, the biggest moon in the Solar System, is larger than the planet Mercury and nearly the size of Mars. Three others are almost as big as Mercury and far bigger than Pluto. If not for Jupiter's glare, these four moons could easily be seen in the night sky, glowing as bright as stars.

Counting out from the Sun, Jupiter is the fifth planet. The first four, including Earth, are mostly rock and metal. Then comes Jupiter, the first of four much larger gas giants. The ninth planet, Pluto, is a ball of ice and dirt smaller than the Moon.

A view of Jupiter from the Voyager 1 spacecraft. One of Jupiter's moons, Io, can be seen in front of the planet.

Why does Jupiter take up 1,300 times as much space, when it is only 318 times as massive? Jupiter's gases, though tightly compressed, are still just one-fourth as dense as rocky Earth.

HOW JUPITER FORMED

Five billion years ago, the Sun and the planets did not exist. There was only a big cloud of spinning gas and dust called a nebula. Gravity pulled part of the nebula together into a ball, which became the Sun. The more matter it collected, the stronger its gravity became. Eventually, the gas in the center of the Sun was pressed together so tightly that hydrogen atoms began fusing to form helium. A huge amount of energy was sent out in the process. This nuclear reaction continues to this day. It provides all the heat and light we get from the Sun.

The four gas giants—Jupiter, Saturn, Uranus, and Neptune—are all far bigger than Earth.

Sizewise

How much bigger is Jupiter than Earth? That depends on how you're measuring. By diameter, it is eleven times bigger, which means it would take eleven Earths to stretch from one side of Jupiter to the other. By total mass—how much matter it contains—Jupiter is 318 times bigger. By volume—how much space it takes up—Jupiter is 1,300 times bigger.

Jupiter contains two and a half times as much matter as all the other planets combined. It is so much bigger than Earth that if the two planets were placed side by side, Earth would look like one of Jupiter's moons.

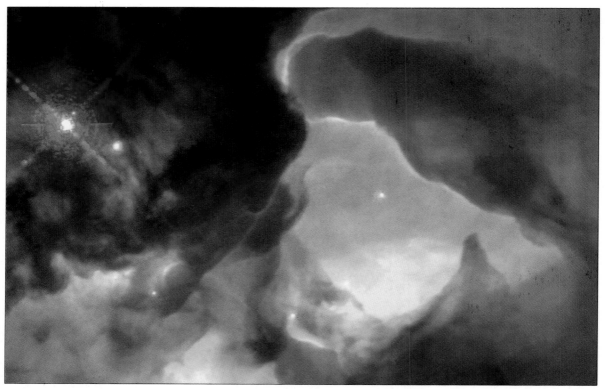

The Sun and planets formed from a giant cloud of gas and dust called a nebula. Many nebulas still exist, including the Lagoon Nebula, photographed here by the Hubble Space Telescope. The red spots are stars that have already formed.

The gas and dust still spinning around the Sun continued to clump together. These balls became the planets. As the growing planets circled the Sun, they swept up more dust and gas from the nebula. But they did not all gather the same amount.

Closer to the Sun, it was hot enough to boil water. The only solid particles were specks of rock and metal. Everything else was gas. The first four planets—Mercury, Venus, Earth, and Mars—formed from these particles, so they are mostly rock and metal.

Farther from the Sun, it was darker and colder, so cold that water, ammonia, and methane gas froze into snowflakes. This snow was ten times more plentiful than the dust that formed the inner planets. Four outer planets—Jupiter, Saturn, Uranus, and Neptune—gathered up these frozen particles and grew far larger than the four inner planets.

But Jupiter did not stop there. When a planet becomes fifteen times more massive than Earth, its gravity is strong enough to attract and hold gases such as helium and hydrogen. Once Jupiter grew this massive, it began sweeping up these gases, which composed 98 percent of the solar nebula. Because it now took in everything in its path, just as the Sun had, its chemical makeup resembled the Sun.

The Solar System was created when a spinning nebula flattened into a disklike shape. The Sun formed in the center, while the rings of dust and gas around it became the planets. The blue lines in this drawing show gases being blown out the nebula.

Out past Jupiter, the nebula was thinner, so Saturn did not collect nearly as much helium and hydrogen as Jupiter. Still farther out, Uranus and Neptune grew so slowly that by the time they became big enough to attract hydrogen and helium, most of these gases had blown away. So while all four are much bigger than Earth, Jupiter is by far the biggest of these gas giants.

INSIDE JUPITER

On Earth, the two lightest gases in the universe, hydrogen and helium, are used to fill blimps and balloons. On Jupiter, these elements take many different forms and are divided into several different layers.

In the thinnest, outermost layer of Jupiter's atmosphere, hydrogen and helium are gases, like on Earth. The layer below the atmosphere is a giant ocean made of liquid hydrogen. Unlike Earth, however, there are no islands or continents in this ocean. Also unlike Earth, there is no clear boundary between the atmosphere and the ocean. The gases just become gradually thicker and thicker until they become liquid.

In an even deeper layer, with the weight of the entire planet pressing down, the pressure is 10 million times greater than in Earth's atmosphere. The hydrogen and helium in this layer are so compressed that they behave like liquid metal and conduct electricity.

Beneath this layer, in the center of Jupiter, is a core of molten rock and iron. It contains only a small fraction of Jupiter's total matter, but it is still fifteen times more massive than Earth.

BULLY OR BUDDY?

With its great mass and gravity, Jupiter dominates the rest of the planets in the Solar System. Jupiter stunted the growth of its neighbor Mars by sweeping the area clean of rocks that Mars would have oth-

In the center of Jupiter is a core of liquid rock. Around the core is a thick layer of hydrogen compressed so tightly it behaves like liquid metal (shown here as dark blue). Around that is a thick layer of liquid hydrogen (lighter blue). A thin outer layer of gas forms Jupiter's atmosphere, which contains hydrogen (pale blue), water (white), ammonium hydrosulphide (brown) and ammonia (red).

erwise collected. Jupiter's biggest victim, though, is the missing planet between Mars and Jupiter that never had a chance to form.

Given the spacing of the planets, one would expect to find another planet between Mars and Jupiter. Instead, this region contains nothing but a bunch of flying rubble called asteroids. Asteroids are chunks of rock and metal much smaller than planets. These asteroids might have formed a planet. But the constant tugging from Jupiter overpowered the efforts of their own gravity to pull them into a ball.

The asteroids orbit the Sun in separate rings or belts, with clear gaps between them. These empty spaces represent the areas most vulnerable to tugs from Jupiter's gravity. The asteroids that once occupied these spaces always reached a particular spot in their orbit just as Jupiter arrived at a similar spot in its orbit. Each time they reached this position, Jupiter's gravity tugged the asteroid out a little farther. Eventually it was pulled out of its orbit and flung into space, much like a series of small pushes can make you go very high on a swing. Some of these wayward asteroids may have been pulled into orbit around Jupiter and become its outer moons.

Jupiter Fact Sheet

Distance from Sun: 483,800,000 miles (778,600,000 km)
Diameter at equator: 88,846 miles (142,984 km)
Mean surface temperature: -166 degrees F (-110° C)
Surface gravity: 2.36 times Earth's gravity
Period of revolution (year): 11.86 Earth years
Period of rotation (day): 9.9 hours
Number of moons: 28

YOUR WEIGHT ON JUPITER

Your weight depends on two things: how much of you there is—your mass—and where you are. On another planet, your mass would be the same, but your weight would change. That's because your weight depends on how strongly that planet's gravity tugs on you. The more mass a planet has, the stronger its gravity. Jupiter, being the Solar System's most massive planet, thus has the strongest gravity, so you would weigh more there than on any other planet.

If you were standing on Jupiter, you would weigh two and a half times as much as you weigh on Earth. But you could never stand on Jupiter, because it doesn't have a firm surface. As you descend into Jupiter's atmosphere, the pressure increases until the gas becomes liquid. There is nothing solid on Jupiter at all.

Between Mars and Jupiter is a belt of large rocks called asteroids. Some of Jupiter's moons may have started out as asteroids that fell into orbit around Jupiter.

But Jupiter is not just a planetary bully. Some astronomers think life could not have survived on Earth without a giant planet like Jupiter around to sweep up dangerous asteroids and comets with its gravitational vacuum cleaner. A single asteroid or comet impact is believed to have killed off the dinosaurs, so you can imagine how dangerous it would be to have such destruction raining down on us regularly. Life in other solar systems may require a similar combination of a small planet like Earth and a big-brother planet like Jupiter to protect it from flying debris.

The Greek god Zeus and the Roman god Jupiter could never fight a battle. Why? They are the same god. The Romans were such great admirers of Greek culture that after conquering Greece, they adopted the Greeks' religion. So the ancient Greeks and Romans both had the same gods, but with different names. The king of the gods was named Zeus in Greek, while the Romans called him Jupiter. Astronomers today use both sets of names. The planet Jupiter is known by its Roman name, while its moons take their names from Greek myths. But they are all characters in the same story.

The planet Jupiter was named for the king of the Roman gods because it moves across the sky from constellation to constellation, like a king checking on his subjects.

2

OBSERVING JUPITER

If you stand in the middle of a room and twirl around, it will look like the room is spinning around you. The same thing happens with the Earth and stars, only much slower. Since you cannot feel the ground moving, people long ago naturally assumed that Earth was the center of the Universe and that everything revolved around it. As the Sun and stars rise in the east and set in the west, it looks like they are traveling in a giant circle around Earth. Actually, it is Earth that is spinning, not the stars.

Yet certain "stars" do not follow the orderly march around Earth. Instead of staying in neat patterns and constellations, these vagabond stars wander about the sky. Sometimes they surge ahead of the stars around them, and sometimes they lag behind. These stars are actually planets. But without a telescope, their strange motion is the only thing that sets them apart from stars.

The Polish astronomer Nicolaus Copernicus was the first person to explain why the planets do not move with the stars: Earth and the other planets are all in constant motion around the Sun, so they line up in a different place against the backdrop of the sky each night.

In 1610, an Italian astronomer, Galileo Galilei, produced strong evidence that Copernicus was right. Galileo was one of the first people to look at Jupiter through a telescope. Galileo's lenses were not even as good as a pair of modern binoculars. But when he trained his homemade telescope on Jupiter, he saw something astounding: Jupiter had moons! And not one moon, but four. As he watched these

Unlike the stars, which appear to stay in a fixed pattern, Jupiter and the other planets seem to drift around in the sky. As the planets move around the Sun, we view them from a different direction. So they line up against a different background each night.

four tiny spots of light journey around Jupiter, he knew he had found proof that not everything in the heavens revolved around Earth.

Religious leaders of the day were not pleased with the theories of this upstart. They insisted that Earth was the center of the Universe. The Catholic Church forced Galileo to publicly retract his views, and he lived out his life under house arrest. But the church could not keep other people from looking through telescopes. Science marched on without its approval.

Astronomer Nicolaus Copernicus changed the way people saw Earth's place in the Universe. Previously, most people thought that the Sun, stars, and planets all revolved around Earth. But Copernicus realized that Earth moves around the Sun.

As telescopes continued to improve, they revealed new features on Jupiter. In 1630, stripes were discovered, and in 1664, people saw spots on Jupiter. As they watched these spots move across the planet's surface, they realized Jupiter was spinning, making a complete rotation every ten hours.

The Great Red Spot was first sighted in 1665 by Italian astronomer Giovanni Domenico Cassini. This giant storm, still visible

The great Italian astronomer Galileo Galilei wrote several books supporting Copernicus's view that Earth was not the center of the Universe. For challenging established ideas, he was imprisoned, and some of his books were burned.

WHO SAW THEM FIRST

Jupiter's four largest moons were the first objects ever seen orbiting another planet. This group of bright satellites is named for Galileo. But no one is sure if he was really the first person to see them.

Galileo published a book in March 1610 in which he described seeing the moons on January 7. A year later, a German astronomer, Simon Marius, announced that he had seen them first. In 1614, Marius published his own book that dated his first recorded observations to December 29, 1609. At the time, Italy and Germany used different calendar systems. Galileo pointed out that December 29, 1609 in Marius's calendar was actually January 8, 1610 in his. So Galileo had still beaten him by a day. Marius claimed he had actually seen the moons a month before he had written anything down, so he was still first. But Galileo charged that Marius had simply copied his observations from Galileo's book.

Traditionally, newfound objects in the sky are named by their discoverer. Galileo and Marius each proposed names for the moons. Galileo wanted to name them after the rich family he worked for, the Medicis. Marius suggested naming them after characters from the myth of Jupiter. Since no one could prove who actually discovered the moons, for many years they went without names. They were simply called Jupiter I, II, III, and IV.

Historians still disagree over who saw the moons first, but their names have been settled. As a group, the four moons are known as the Galilean moons, after Galileo. But individually, they are known by the names suggested by Marius—Callisto, Ganymede, Europa, and Io.

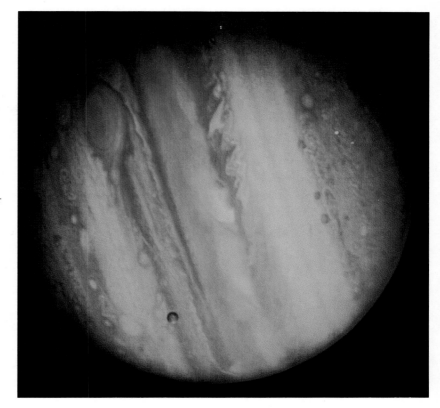

All the features visible on Jupiter are clouds. The Great Red Spot is a gigantic storm that has been raging for hundreds of years. The stripes are bands of clouds that move around the plant. There is no land below the clouds, only liquid hydrogen.

today, has existed for more than three centuries. Cassini was also the first person to figure out the distance between planets. Jupiter was found to be five times farther from the Sun than Earth is. With this knowledge, people could calculate Jupiter's size. Jupiter's mass could also be determined by measuring the effect of Jupiter's gravity on its moons.

In 1892, a fifth moon of Jupiter was discovered by E. E. Barnard in California. This moon was much smaller than the Galilean moons and orbited much closer to Jupiter. It was so dim that Barnard was only able to see it by covering up Jupiter in his telescope. Barnard named this moon Amalthea, in honor of the mythical goat whose milk the god Jupiter had drunk as a baby. Amalthea was the last moon discovered by looking through a telescope. All the moons discovered

Gerard Kuiper calculated the amount of methane and ammonia in Jupiter's atmosphere in the 1950s. Kuiper also predicted that thousands of smaller objects circle beyond the major planets. The ninth planet, Pluto, is thought to be one of these objects.

since then have been located using long-exposure photographs, which gather more light than the eye can see.

Other new technology helped scientists detect specific chemicals in Jupiter's atmosphere. Spectrometers, for instance, measure the frequency of light absorbed and emitted by different elements. In the 1930s, spectrometers detected methane and ammonia in Jupiter's atmosphere. By the 1950s, astronomer Gerard Kuiper had figured out their proportions, which together total less than one percent.

By then, the world was poised for a major advance in astronomical observation—launching satellites into space. The Soviet Union sent the first artificial satellite into orbit, *Sputnik I*, in 1957. The United States rushed to catch up, and the space race had begun. Probes were launched to survey Earth, the Moon, Mars, and Venus in the 1960s. Finally, it was Jupiter's turn.

The trip to Jupiter posed dangers that no other mission had faced. For one thing, no one knew whether a spacecraft could survive the intense radiation from ions trapped in Jupiter's magnetic field. It is a thousand times the lethal dose for a human being. Also, no spacecraft had ever crossed the Asteroid Belt, and some people worried a probe might be shattered by flying rocks. So, before a major mission was attempted, NASA sent out a pair of relatively simple and inexpensive probes as a test.

Called *Pioneer 10* and *11*, these probes were launched in 1972 and 1973. They made it through the Asteroid Belt without a problem and handled the intense radiation around Jupiter with ease. The *Pioneer* probes measured infrared emissions, which provided the first accurate readings of the temperature at different levels of the atmosphere. They also gathered data about Jupiter's magnetic field. This showed that the field is generated by electricity flowing through the metallic hydrogen around the planet's core, not in the core itself, as it is on Earth.

Is Anybody out There?

The *Pioneer* probes flew past Jupiter, Saturn, and out into space. Since they were the first humanmade objects to ever leave the Solar System, NASA included a gold-plated plaque with a message to any aliens who might find it. Each plaque contains a diagram showing a man, a woman, and the location of the Earth. The man is raising his hand in a friendly wave. Since they will not reach the nearest star for several million years, it could be a while before anyone waves back.

Although only a few of the close-up pictures taken from the probes were better than those taken from Earth, the *Pioneers* had achieved their main mission, which was simply to prove a voyage to Jupiter could succeed.

The *Pioneer* probes were followed in 1977 by two more advanced spacecraft, *Voyager 1* and *2*. The *Voyagers* had cameras that took pictures sharper than those produced by the best telescopes on Earth. Instead of *Pioneer*'s few dozen close-up pictures, the *Voyagers* took more than 30,000, including several frames of video.

These probes discovered three new moons of Jupiter and sent back images of several active volcanoes erupting on Jupiter's moon

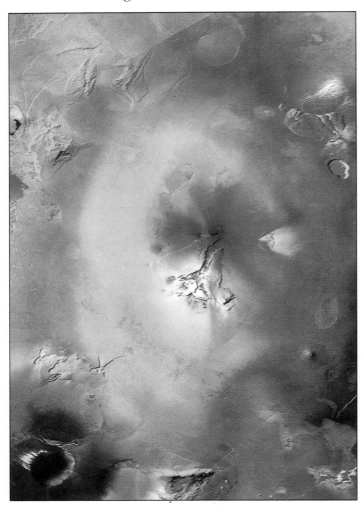

The Voyager 1 *spacecraft took this picture of an enormous volcano named Loki erupting on Jupiter's moon Io.*

Io. Even more surprising, they found a faint ring of dust around Jupiter. Unlike Saturn's bold rings, which can easily be seen from Earth, Jupiter's ring is so dim that it is nearly invisible.

The *Voyager* missions were among the most valuable space expeditions in history. For the next two decades, almost everything known about Jupiter and its moons came from *Voyager* data.

The next generation of planetary exploration was the *Galileo* spacecraft. It was launched from the space shuttle *Atlantis* in 1989 and arrived at Jupiter in December 1995. *Galileo*'s most dramatic experiment was dropping a probe into Jupiter's atmosphere. It entered Jupiter's atmosphere at more than 100,000 miles per hour (170,000 km/hr). Friction with the atmosphere heated its nose to 10,000 degrees Celsius (18,000° F), but a heat shield protected the probe. As the shield melted in the intense heat, the molten drops blew off, taking much of the heat with them. When the probe slowed to 2,000 miles per hour (3,000 km/hr), it ejected its shield and opened a parachute. Seven instruments then began recording the temperature, pressure, density, and makeup of the atmosphere. The probe sent back data by radio as it plunged 125 miles (200 km) into the atmosphere, until it was crushed by the ever-mounting pressure.

The probe turned up a few surprises. One was that helium and hydrogen were found in the same proportion as in the original solar nebula. Scientists had expected to find that some of the heavier helium atoms had sunk deep into the planet. But apparently Jupiter has been churning with internal heat for billions of years, keeping things evenly mixed. The probe also detected less ammonia and water than expected. But it may have simply fallen in between the clouds that contained them.

The probe found a disappointing lack of the larger molecules that form in the presence of ammonia, water, and ultraviolet light. Some people have speculated that life could evolve from molecules that form this way. Carl Sagan and others have imagined balloonlike

An artist's view of the Galileo *probe as it descends into Jupiter's atmosphere by parachute. The* Galileo *spacecraft that launched it is shown in orbit above.*

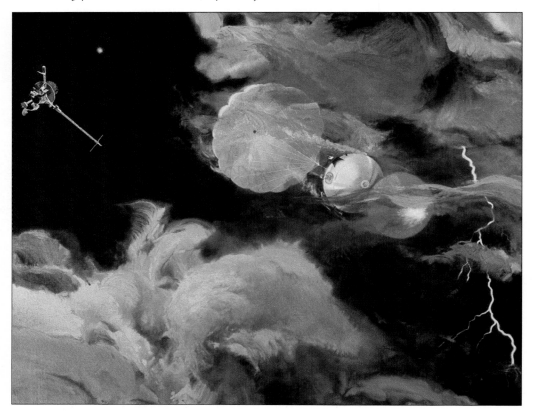

"floaters" and winged "hunters" that could theoretically live in Jupiter's atmosphere and even provide the colors of some of its clouds.

The most likely place to search for life, however, is not in Jupiter's atmosphere, but on its moons. There, liquid water, heat, and chemical nutrients—the keys to life on Earth—are all available. The *Galileo* spacecraft flew near all four Galilean moons and found signs of liquid water on three of them. Whether anything is living in this water will be left to future missions to discover.

The Galileo *spacecraft took six years to reach Jupiter. Once in orbit around the planet, its mission was extended for several years more than originally planned, allowing it to continue making new discoveries.*

3

JOVIAN WEATHER

From Earth you can see stripes on the surface of Jupiter using a simple telescope. They change shape constantly, and over time, they change color as well. These stripes are belts of clouds that stream around the planet at up to 400 miles per hour (600 km/hr). The highest clouds are made of ammonia. Lower down, they are composed of water.

Jupiter's clouds come in many different colors, including brown, orange-red, pale yellow, gray, white, and bluish white. The clouds' colors come from different chemicals, such as sulfur and phosphorous, that are present. The color of a cloud indicates how high up it is. Orange-red clouds are the highest. Next come white clouds. Most brown and grey clouds are lower still. Bluish clouds are the lowest and are normally visible only through holes in the clouds above.

Each colored ring of clouds moves in the opposite direction to the one next to it. Along the borders of these bands, clouds slide against each other, forming curls and eddies. Occasionally, these spinning tendrils curl all the way into a ball that breaks free from the rows of clouds. These spinning balls form giant storms that can last for many years.

THE GREAT RED SPOT

Jupiter's biggest storm is twice as wide as Earth. Known as the Great Red Spot, it measures 7,750 miles (12,400 km) from north to south and

The entire surface of Jupiter is covered with swirling clouds whose borders twist and intertwine as they rub against each other.

These photographs from the Hubble Space Telescope show how Jupiter's Great Red Spot changed from 1992 to 1999. The size, shape, and color of the gigantic storm have been changing during the more than three hundred years it has been observed.

stretches 14,400 miles (23,000 km) from east to west. It swirls counter-clockwise, with the winds on its outer edge reaching 250 miles per hour (400 km/hr).

The Great Red Spot is the oldest known storm in the Solar System. Though there were periods when it was not observed, it is believed to have simply changed color, not disappeared.

But how can a single storm last more than three centuries? Before astronomers realized Jupiter had no solid surface, some thought the Great Red Spot might be swirling around a giant mountain or some other obstacle below the clouds. Another theory was that it might be driven by heat flowing up from a particular area. By comparing recent observations with historic drawings, however, NASA's Reta Beebe concluded that the Great Red Spot is not anchored to any one place. Instead, it wanders freely. From 1910 to 1940, the storm moved faster than the atmosphere around it. Then, for the next twenty years, it

This photo of Jupiter's clouds taken by the Galileo *spacecraft has been computer enhanced to include invisible infrared light. "False color" images such as this include hues that were added to the photograph to make a certain feature sharper or more visible.*

drifted slower than the nearby clouds. In the years since then, it has changed speeds several times. But no matter how much it wanders east or west, it never moves north or south.

The Great Red Spot is sandwiched between two belts of wind moving in opposite directions. Just north of it, near the equator, is Jupiter's strongest westward-moving belt. On its south is a slower, eastward-moving belt. These two winds help keep the immense spot spinning. The storm also gains energy by swallowing up smaller storms and eddies. And, like all the weather on Jupiter, it is partially driven by heat streaming up from below. Whatever is driving the Great Red Spot, it is not expected to stop spinning in our lifetimes.

OTHER STORMS

Three white ovals circle Jupiter just south of the Great Red Spot. They began forming in 1939, when three brown gaps appeared in a band of white clouds. The sections of the broken white band gradually formed oval-shaped storm systems that spin counterclockwise, like the Great Red Spot. Also like that giant storm, the white ovals sometimes swallow up smaller eddies, whose energy helps keep them spinning. Because the white ovals move around the planet faster than the Red Spot, each one passes it every two and a half years.

A ring of dark clouds, called barges, circles Jupiter near its north pole. The *Voyager* mission observed four of these long dark smudges, but the number changes. Because they move at different speeds, a large barge will occasionally catch up to a smaller one and swallow it.

Unlike the Great Red Spot, which bulges upward, the barges are actually funnel-shaped holes in the clouds around them. So while the Red Spot gets its color from chemicals pushed up from below, the barges are dark because they suck shiny, white ammonia clouds down to lower, warmer regions. There the clouds melt and disappear, leaving behind only a brown haze.

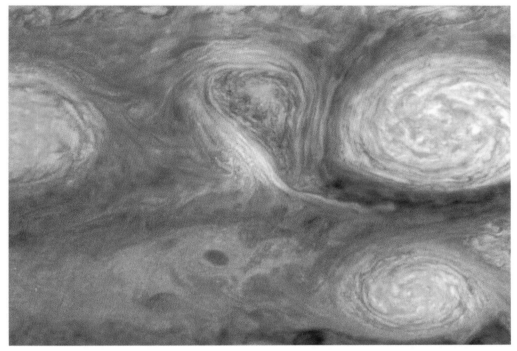

The two light blue ovals in this picture are storms spinning counterclockwise. The brown, pear-shaped storm near them is spinning in the opposite direction. In 1998, after this picture was taken by the Galileo *spacecraft, they merged into a single, larger oval.*

LIGHTNING

The highest layer of Jupiter's atmosphere contains almost no water. At that height, it is so cold that any water would freeze and fall as snow. Lower down, though, beneath the clouds of ammonia, water forms thunderclouds similar to those on Earth. These storm clouds generate huge bolts of lightning. When the *Voyager* probes flew around the dark side of Jupiter—the side that faces away from the Sun—they photographed several lightning storms in action. Though Jupiter's lightning is seen through a veil of clouds, it is still much brighter than the lightning on Earth.

A view of Earth's stunning aurora borealis, or northern lights, as seen from Anchorage, Alaska. The glowing lights in Jupiter's atmosphere are believed to be even more spectacular.

AURORAS

From time to time on Earth, odd glowing shapes appear in the skies near the North and South Poles. These rippling lines are called auroras. In the United States and Canada, people also call them northern lights. Auroras occur when particles called ions—atoms that carry an electric charge—stream out from the Sun and hit the atoms in our atmosphere, leaving a glowing trail. Earth's magnetic field channels these ions toward the poles, which is why auroras can be seen best in the far north and south.

Jupiter has an even stronger magnetic field than Earth, and it too channels passing ions toward its poles. However, in addition to ions from the Sun, Jupiter also receives a spray of ions from its volcanic moon, Io. Io's nonstop volcanoes constantly shoot out gases and particles. Since Io's gravity is so weak, many particles never settle back onto the moon's surface. Instead, they trail behind Io in space and steadily rain down on Jupiter. When they strike the gases in Jupiter's atmosphere, glowing auroras appear.

The blue lines in this picture represent Jupiter's invisible magnetic field. As electrically charged particles stream toward Jupiter from the Sun (arrows), the planet's magnetic field deflects them toward the poles.

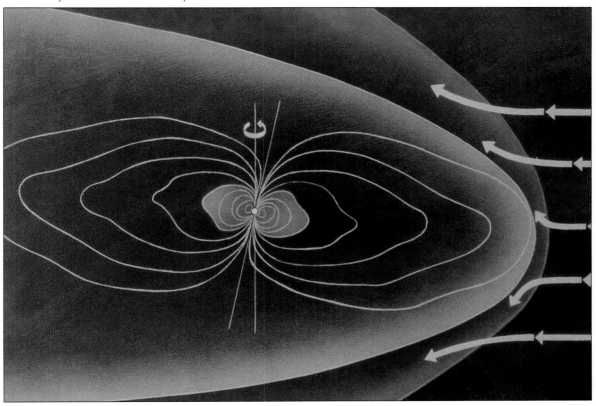

4

MOONS, MOONS, MOONS

No one can ever know exactly how many moons a planet has. All anyone can say for sure is the number that have been discovered *so far*. This point was made dramatically clear in January 2001, when astronomers at the University of Hawaii announced they had just discovered eleven more moons orbiting Jupiter, the largest number of moons ever discovered at one time. This record discovery raised Jupiter's total number of known satellites from seventeen to twenty-eight.

The moons of Jupiter fall into three basic groups: four small inner moons, four giant Galilean moons, and twenty tiny outer moons. Though the origin of the inner moons is uncertain, the Galilean moons probably formed at the same time as Jupiter. Both groups move in the same direction in which Jupiter spins. They orbit in nearly perfect circles directly over Jupiter's equator.

The outer moons, in contrast, have extremely irregular orbits. They travel in long ovals that are tilted in relation to the direction of Jupiter's rotation. Fourteen of them actually move opposite to the direction that Jupiter turns. This is called a retrograde orbit. Their odd-shaped orbits indicate that these moons did not form from the same spinning blob of gas and dust as Jupiter. Instead they originated as asteroids that were captured by Jupiter's gravity.

What could slow down a passing asteroid enough to make it fall into orbit and become a moon? Long ago, Jupiter's atmosphere might have extended much farther into space than it does today. If a passing

This view of Jupiter and its four largest moons was made by combining several photographs. Above the surface of cratered Callisto, from left to right, are Jupiter, Io, Europa, and Ganymede.

asteroid hit this bloated atmosphere, friction would slow and heat it, like the fiery meteors that hit Earth's atmosphere and leave a brief, glowing trail. Just as meteors sometimes explode in giant fireballs, a passing asteroid could have been blown apart by friction with Jupiter's atmosphere. Since many of Jupiter's outer moons travel in clusters with similar orbits, they quite likely started as larger asteroids that broke apart. These asteroids exploded because of friction. Or they were pulled apart by Jupiter's gravity.

The Galilean moons of Jupiter, as photographed by Voyager 1. *At upper left, covered with a spray of orange volcanic ash, is Io. At upper right is Europa, whose brown streaks may show where dirty water leaked onto its icy surface. At lower left is Ganymede, the largest moon in the Solar System. At lower right is crater-scarred Callisto.*

How to Find a Moon

In pictures taken through a telescope, moons look just like stars. To find a moving object such as a moon, asteroid, or comet, two or more pictures must be compared. The stars stay in the same place in each picture, while a moon will have moved to a new spot. Comparing pictures is a slow, difficult process. Staring at pictures of thousands of white spots, it is very difficult to tell if one is no longer in the same place. And one must look at thousands and thousands of pictures to find one moon. Galileo found four moons of Jupiter just by looking through his telescope, but it took S. B. Nicholson, the next person to find four moons of Jupiter, more than 37 years of work, from 1914 to 1951. Today, computers make this painstaking procedure quite a bit easier.

THE GALILEAN MOONS

Though more moons may yet be discovered circling Jupiter, none will match the size or complexity of the Galilean moons. These four huge moons are all Solar System record-holders. One ranks first in craters; another is the system's largest moon; a third is the smoothest, shiniest moon; and a fourth has the hottest and most volcanic surface in the Solar System. Each of these planet-sized moons is fascinating in its own right.

Callisto

This satellite is the outermost of the Galilean moons, orbiting 1,170,000 miles (1,883,000 km) from Jupiter. With a radius of 3,000

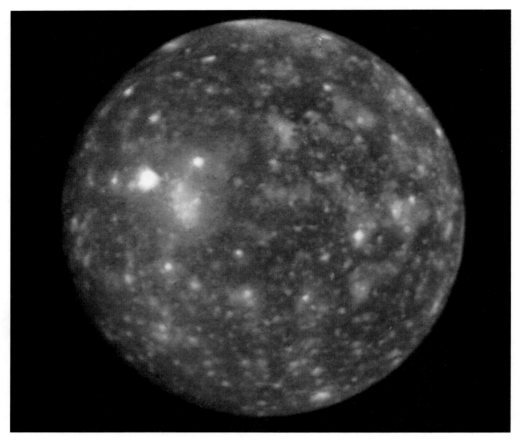

Callisto is dotted with white spots where meteorites punched through its thin crust to expose the cleaner ice below.

miles (4,800 km), Callisto is almost exactly the same size as the planet Mercury. It looks like Mercury, too. Both have heavily cratered surfaces. But Mercury has a few more flat spots, so Callisto gets the nod as the most cratered place in the Solar System. Unlike Earth's moon, which had many of its craters filled in by flowing lava, Callisto has never experienced volcanic activity. Every pockmark made by a meteor impact remains.

One strange feature of Callisto's craters is that they are all relatively small—less than 30 miles (50 km) across. Many craters on other moons, including Earth's, are far larger. One explanation could be that objects approaching Callisto were ripped apart by Jupiter's intense gravity, so that only small pieces of cosmic debris struck the moon. It is also possible that larger craters on Callisto simply collapsed, because its outer layer is composed not of rock, but of fragile sheets of ice.

Because its surface is covered with dust from meteor impacts, Callisto resembles a reddish, rocky asteroid. Callisto's youngest craters are the whitest, because the collisions that created them exposed the fresher ice or slush below. The oldest craters are darker because they have been covered by dirt and dust, just as freshly fallen

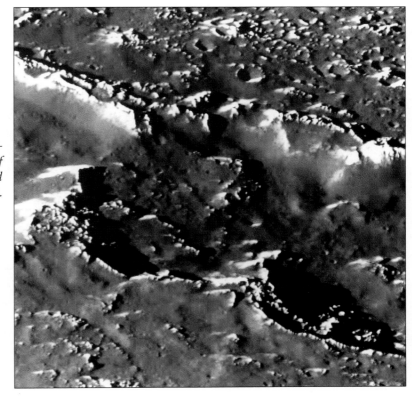

A chain of craters was created when a string of comet fragments slammed into Callisto.

snow starts out white but gets darker and dirtier the longer it stays on the ground.

Callisto's most remarkable feature is a giant bull's-eye pattern that covers most of one side. Called Valhalla, this pattern resembles the rings that ripple outward when a pebble is dropped into water. In the middle is a white patch about 375 miles (600 km) across. Valhalla's widest circles extend about 750 miles (1,200 km) from the center, and partial rings go out about 1,200 miles (2,000 km). With a total width of 2,500 miles (4,000 km), this is the largest multiringed basin in the Solar System, ahead of a 2,000-mile (3,200-km) ringed basin on our moon. Another bulls-eye basin on Callisto named Asgard contains circles with a radius of 500 miles (800 km). A third basin is about half that wide.

Ganymede is covered with light-colored bands curving across a darker surface. The dark areas are plates of older crust that stretched and moved to reveal the lighter areas beneath.

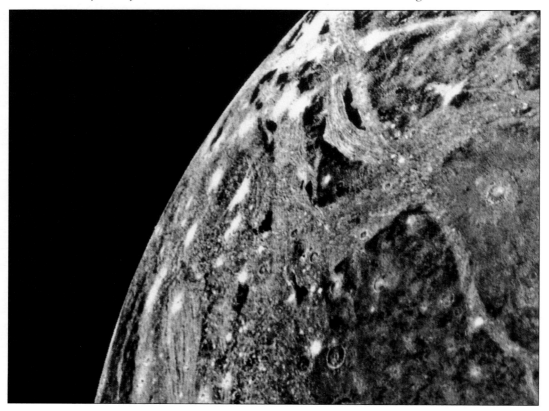

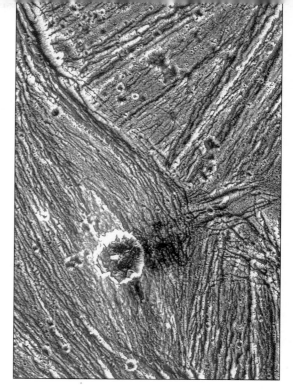

The light areas on Ganymede are covered by rows of grooves and ridges. The grooves are believed to have formed when the moon expanded millions of years ago, causing cracks to appear in its rigid surface as it stretched apart.

Although these circular ridges were clearly formed by a major impact, they are probably not actual ripples of waves, since Callisto is solid, not liquid. Most likely, an impact caused cracking and settling in Callisto's thin crust.

Callisto was long considered geologically inactive, because it has no volcanoes or weather. This means that nothing erodes or fills in craters, even after million of years. But the *Galileo* mission did turn up an unexpected finding. Callisto has its own thin atmosphere, made of hydrogen and carbon dioxide. Another surprise was that under its frozen surface, some water may be sloshing around.

Ganymede

The next closest moon to Jupiter is the giant Ganymede, the Solar System's largest moon. Twice the diameter of Pluto, Ganymede is bigger than Mercury and nearly the size of Mars. Its rocky core is the size of Earth's moon. But this core is covered with an equal mass of ice.

The surface of Ganymede is a blend of light and dark patches. The dark areas are filled with ancient craters, like the surface of Callisto. But between these dark patches stretch winding, curved areas of lighter-colored material, where sections of the ancient crust slid apart, allowing new material to ooze up from below and fill in the gaps. These white patches, which cover about 60 percent of Ganymede's surface, are crisscrossed with grooves and cracks. Though the light patches are younger than the darker ones, they are still older than any features on Earth, as Ganymede's plates stopped moving long ago.

Ganymede is more geologically active than Callisto. This may be due to its greater internal heat. Since Ganymede is closer to Jupiter, it receives three times more heat from Jupiter than Callisto. It also has more rock than Callisto, which holds heat better. Overall, though, Ganymede resembles Callisto much more than Europa and Io, the Galilean moons even closer to Jupiter.

Europa

While Callisto and Ganymede are both about half ice, Europa and Io are mostly rock and iron. They are smaller, too, because they formed at higher temperatures and thus did not collect a thick layer of ice around their rocky bodies. Still, Europa is about 15 percent water by volume. This water forms a thin layer around a big ball of rock with an iron core.

Since Europa is warmer than Ganymede and Callisto, most of its water is liquid. Beneath a top layer of ice lies an ocean so big that it contains more water than all of Earth's oceans combined. This vast sea is considered by some to be the best place in the Solar System to look for extraterrestrial life.

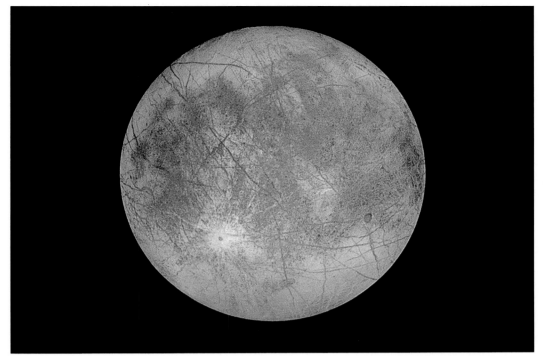

Europa, the shiniest moon in the Solar System, is covered with ice. The dark streaks on the surface show where dirty water appears to have leaked up through cracks in the icy crust.

*This picture taken by the **Galileo** spacecraft shows the ridges on Europa's icy surface in false color. The ridges (orange) probably formed when mineral-filled water oozed up through cracks in the moon's icy surface (blue).*

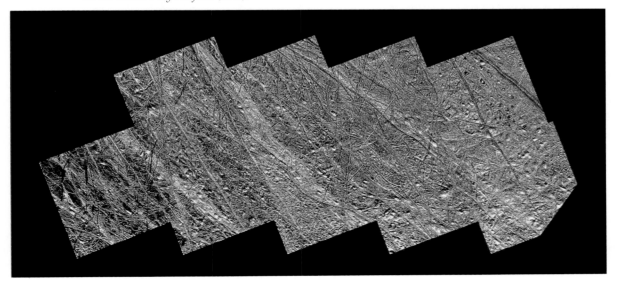

Because Europa's surface is a single vast sheet of shiny white ice, it is extremely reflective. This makes it the brightest, shiniest moon in the Solar System. This remarkably smooth outer covering is almost completely unscarred by craters. In fact, the *Voyager* mission found a total of just twelve craters on Europa's entire surface. Though Europa has been struck by as many meteors as pockmarked Callisto, its craters have been filled in with water, leaving an even layer of ice.

The only thing that mars Europa's white surface is a network of dark, crisscrossed lines and some brown blotchy spots. Both are probably areas of dirty ice, where dark material oozed up from below or where ice evaporated, leaving dirt and minerals behind. Cracks and ridges border these dark lines, but none rises high enough to break the overall smoothness. *Voyager 2* could not find a single ridge that rose even a half mile (1 km) above Europa's surface.

Io

This moon is the most volcanic spot in the Solar System. It has the hottest surface except for the Sun. It has an iron core surrounded by a thick layer of molten rock, which is barely covered by a thin crust of solid rock.

Io is the Galilean moon closest to Jupiter, but this does not account for its heat and volcanoes. Jupiter's great gravity pulls on the side of Io facing the planet so strongly that the moon actually stretches. Ganymede's gravity, on the other hand, rocks Io from side to side as it passes, making Io's insides slosh around like water in a bucket. All this stretching and squeezing generates enough heat to melt the rock inside Io.

One hundred twenty cubic miles (500 cubic km) of lava flow onto Io's surface each year—one hundred times the amount produced by all the volcanoes on Earth. Though Io is about the same size as Earth's moon and is made of similar material, it releases one hundred times

Two giant volcanic plumes on Io can be seen erupting in this picture. On the right edge of the picture, a plume shoots powder 75 miles (120 km) above the surface. In the center, seen from directly above, is a volcano named Prometheus. It is ringed by a dark shadow. Prometheus has been erupting for more than eighteen years.

more heat than our moon because of its many volcanoes. About eighty active volcanoes have been found on Io.

The hot magma inside Io gushes onto the surface in dozens of places at once. The *Voyager* probes observed more than 200 calderas, or pools of lava, that were more than 12 miles (20 km) wide. There are only fifteen calderas that large on Earth. One lake of liquid rock and sulfur near a volcano named Loki measures 125 miles (200 km) across. It is large enough to swallow the entire island chain of Hawaii.

Even more dramatic than the lakes of liquid lava are Io's volcanic "plumes." They shoot umbrella-shaped showers of powder and gas up to 250 miles (400 km) above its surface. The largest plume, named Pele in honor of the Hawaiian volcano goddess (all the features on Io are named for gods of fire or volcanoes), has been erupting constantly

INNER MOONS AND RING

Inside Io's orbit are four much smaller moons and a ring of dust. The largest of these, Amalthea, is an oblong chunk of cratered rock 168 miles (270) km long and 96 miles (155 km) wide. The other three inner moons are much smaller, ranging from 16 to 56 miles (2690 km) long. Their irregular shapes indicate that they may have once been part of an older moon or asteroid that broke apart.

    The two innermost moons orbit just outside the edge of Jupiter's ring. They are considered the ring's shepherds, because their gravity keeps the dust particles from drifting out into space. But there is nothing to stop the dust on the inside of the ring from raining down into Jupiter's atmosphere. Enough dust falls that the entire ring should be gone by now. It remains, however, because the dust is always being replaced. New dust is constantly created from collisions between larger particles that orbit within the ring.

Jupiter's two innermost moons, Metis and Adrastea, are called shepherd satellites, because they tend to contain the particles that make up Jupiter's thin ring.

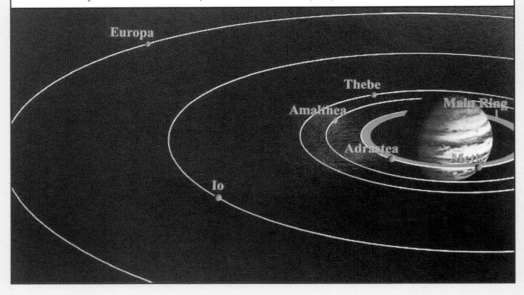

An artist's view of the surface of Io, with Jupiter looming in the sky above. Io is the most volcanic place in the Solar System, partly because it is so close to Jupiter that the planet's gravity causes it to expand.

since it was first spotted by *Voyager* in 1979.

Io's plumes are not made of lava. Instead, they are more like geysers than true volcanoes. On Earth, geysers occur when water comes in contact with magma-heated rock. The water boils and bursts out of a hole in the surface. On Io, however, the geysers are formed not by water but by liquid sulfur or sulfur dioxide. The sulfur is superheated by magma and shoots out like steam from a teakettle. But it instantly freezes as it is ejected, creating a shower of sulfury snow. Because Io's gravity is only one-sixth that of Earth and because there is almost no atmosphere to slow it down, this snow shoots upward for miles before raining down. When it finally settles, it leaves a bright red circle on Io's surface.

Between its volcanoes and plumes, about half an inch (1 cm) of matter is added to Io's surface every year. This constant blanketing explains why not one impact crater remains visible on Io, even though more meteors have crashed into Io than Earth's moon.

STILL SEARCHING

In March 1993, a strangely shaped comet was spotted near Jupiter. It was named Shoemaker-Levy 9, in honor of its discoverers, Eugene and Carolyn Shoemaker and David Levy. The comet's nucleus appeared to be oblong instead of round, with a series of tiny tails jetting out from it. A pair of dusty wings projected from the sides.

Photos from more powerful telescopes revealed it was actually several glowing fragments in a row, which looked like a string of pearls. Astronomers concluded that Shoemaker-Levy 9 had started as a single comet but had been torn apart the year before when it passed within 13,000 miles (21,000 km) of Jupiter. The fragments were on a course that would send them crashing into Jupiter in July 1994.

With a year's warning, astronomers around the world eagerly prepared for the massive collision. It would take place on the side of Jupiter facing away from Earth, but fortunately the *Galileo* spacecraft, then on its way to Jupiter, was in position to monitor the impact. The Hubble Space Telescope also focused on the action.

Over the course of a week, from July 16 to 22, 1994, twenty-one pieces of the comet slammed into Jupiter at 134,000 miles per hour (216,000 km/hr). The energy from each impact was estimated at around 25,000 megatons of TNT. When the comet fragments hit Jupiter's atmosphere, they exploded in huge fireballs. The explosive material then rained down so fast that more superheating occurred. This left huge, dark spots in the clouds that remained visible for months.

Astronomers had hoped that the impacts would kick up matter

Looking like a string of pearls in the sky, fragments of Comet Shoemaker-Levy 9 approach Jupiter. The comet was torn apart by Jupiter's gravity before it plummeted into the planet.

from layers in Jupiter's atmosphere too deep to be seen otherwise. But the comet pieces exploded so quickly that they did not penetrate very far. Though dramatic, the explosions were too violent to be useful to researchers. They had hoped to learn something about Jupiter's atmosphere by observing the collision. But some scientists compared this approach to blowing up a watch in order to see its inner workings.

Gigantic explosions caused by the impact of pieces of Comet Shoemaker-Levy 9 left huge, dark splotches in Jupiter's atmosphere that remained for months.

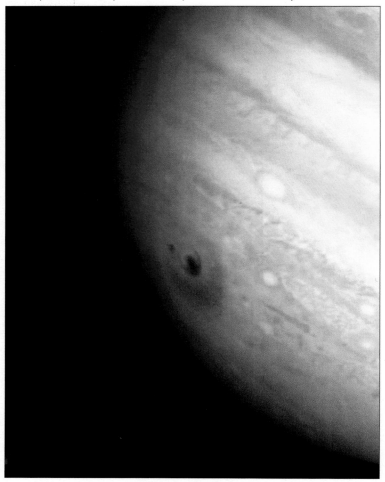

FUTURE RESEARCH

To find out more about Jupiter, more missions will be needed. The *Galileo* probe accessed a single location where there happened to be almost no clouds. Additional probes could provide a much broader picture of Jupiter's atmosphere. One possible plan would be to place an antenna in permanent orbit around Jupiter. Various small probes could use it to relay messages back to Earth.

Better telescopes back on Earth will also allow us to find out more about Jupiter. New cameras were installed on the orbiting Hubble Space Telescope in 1993 and 1997. They offer improved observations, and more upgrades are planned.

An artist's depiction of the immense explosions in Jupiter's atmosphere caused by the impact of pieces of Comet Shoemaker Levy-9. Each explosion released energy equivalent to 250,000 hydrogen bombs.

THE SEARCH FOR LIFE

One of the most exciting areas for future research will be in the search for extraterrestrial life. Although the possibility has not been ruled out, few people expect to find anything living in the gases of Jupiter itself. Jupiter's moons, however, offer some of the best places in our Solar System to look for extraterrestrial life.

The living things of Earth are composed mostly of hydrogen, oxygen, carbon, and nitrogen—four of the most abundant elements in the

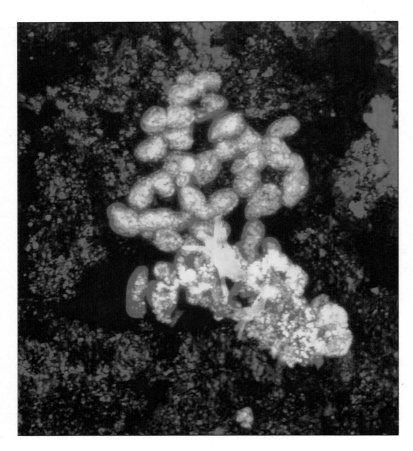

These microscopic bacteria were discovered living on Earth more than half a mile (1,000 m) underground. The discovery of bacteria that can live without sunshine or air raises the possibility that similar life-forms could exist beneath the surface of Europa, Ganymede, or Callisto.

Universe. For life to develop from these building blocks, water, energy, and chemical nutrients must also be present.

The energy source for most life on Earth is the Sun. Plants use this energy to grow, and animals eat plants to release this energy. For a long time, people thought life on other planets must depend on sunlight, too. This made cold, airless planets and moons appear to be among the least likely places to find life.

But in the 1980s, people discovered forms of life on Earth that do not need sunlight to survive. At the bottom of the ocean, in pitch darkness, bacteria were found living entirely on energy and nutrients from volcanic vents in the ocean floor. Bacteria were also found that live deep underground in warm, moist rock. Life-forms such as these that survive in extremely harsh conditions are called extremophiles. Extremophiles have been found in the boiling-hot water in hot springs, and inside the ice in Antarctica.

The discovery of extremophiles raises new hopes that life could be found on Europa, Ganymede, or Callisto. Because of its huge quantity of liquid water, Europa has long been considered a prime place to look for these hardy microbes. Life that arose there could easily have adapted to the cold conditions, just like the extremophiles in Antarctica. If the interiors of Callisto and Ganymede are also warm enough to melt ice, as evidence from the *Galileo* spacecraft indicates, these wet, chemical-rich environments could host life as well.

The only way to find out for sure if life exists on Jupiter's moons will be to send more spacecraft to study them. The exploration of these worlds has barely begun.

GLOSSARY

ammonia a chemical composed of nitrogen and hydrogen

asteroid a rocky body smaller than a planet

element one of the basic chemicals from which everything else is made

comet a ball of ice and dirt that orbits the Sun, sometimes with a tail of gas and dust

conduct allow to flow through

gas a form of matter in which molecules or atoms float freely instead of being connected together, as they are in a solid

geology the study of the processes that shape and change Earth and other planets

gravity the force of attraction between objects; the more massive the object, the stronger the force

magma hot, liquid, underground rock

magnetic field the space around an electric current in which magnetic forces can be felt

mass a measure of the amount of matter something contains

megaton a measure of explosive power equivalent to a million tons of TNT

methane a chemical composed of carbon and hydrogen

nebula a giant cloud of gas and dust in space, from which stars and planets form

orbit the path an object takes around a larger object in space, such as a planet around the Sun.

planet a large ball of matter orbiting a star

pressure a force that squeezes or compresses something

probe a machine that explores and gathers informations

rotate to turn or spin, like a steering wheel or a globe on a stand

satellite an object that orbits a larger object, such as a moon around a planet

solar having to do with the sun

Solar System the Sun and everything orbiting it, including planets, moons, asteroids, and comets

star a giant ball of gas with a nuclear reaction in the center that produces light and heat

Sun the star closest to Earth that provides us with light and heat

TNT an abbreviation for trinitrotoluene, a powerful explosive.

vapor matter in gas form

volume a measure of the amount of space that something takes up

weight a measure of the force of gravity on an object

FIND OUT MORE

BOOKS FOR YOUNG READERS

Bond, Peter. DK Guide to Space: A Photographic Journey through the Universe. New York: Dorling Kindersley, 1999.

Fradin, Dennis Brindell. The Planet Hunters: The Search for Other Worlds. New York: Simon & Schuster, 1997.

Henbest, Nigel. The Planets: Portraits of New Worlds. New York: Viking, 1992

Redfern, Martin. Kingfisher Young People's Book of Space. New York: Kingfisher, 1998.

BIBLIOGRAPHY

I found the following books to be especially helpful in writing this book. They were written by the people who know the most about Jupiter—scientists actively involved in current research.

Beebe, Reta. Jupiter: The Giant Planet. Washington, D.C.: Smithsonian Institution Press, 1997.

Beatty, J. Kelly, Carolyn Collins Petersen, and Andrew Chaikin, eds. The New Solar System. Cambridge, MA: Sky & Telescope/Sky Publishing and Cambridge University Press, 1999.

Rogers, John H. The Giant Planet Jupiter. New York: Cambridge University. Press, 1995.

Rothery, David A. Satellites of the Outer Planets. New York: Oxford University Press, 1992.

WEBSITES

When I wrote this book, the Galileo and Cassini spacecraft were both still orbiting Jupiter and sending back new findings. I relied on the Internet to keep track of their latest discoveries. The following websites are all recommended:

http://galileo.jpl.nasa.gov/
The official NASA homepage for the Galileo mission, with regular updates of the mission's discoveries

http://pds.jpl.nasa.gov/planets/welcome/jupiter.htm
The latest pictures from NASA of Jupiter and its moons

http://www.seds.org/billa/tnp/jupiter.html
Information on Jupiter and its moons with links to other sites.

www.science.nasa.gov
A searchable collection of news about astronomy from NASA

www.space.com
All the latest astronomy-related news

www.spacekids.com
Fun facts and games about space for kids

ABOUT THE AUTHOR

Martin Schwabacher has written more than twenty books for children, including books about asteroids and meteorites, weird rocks, and animals. He writes for websites and exhibitions at the American Museum of Natural History and has contributed to other websites about space. He grew up in Minnesota and currently lives in New York City.

INDEX

Okay, most kids don't need to ask their fathers to show them their mothers at sixteen, right? Mom's out there in the kitchen. Or Mom's out in the car waiting to take you to school. Or Mom's asking if you think she's only made for looking after you and your friends. She's somewhere. Somewhere there's a Mom.

Not in my house. In my house it's Taxi who's in the kitchen, or Taxi who's warming up the Volvo to run me to my cello lesson, or Taxi who's in the conservatorio drinking too much coffee in front of the stereo playing some moronic sliding-saxophone jazz. Frankly, though, we don't seem to bump into each other all that much—just enough, really. I'm not sure I could have stood having anyone else around to fill in all the little spaces in my life that I move through. Taxi is never in those spaces. It's like he knows where I need to go and slips out of the way quietly.

"Brooks is a genuine storyteller, with fine dramatic sense and excellent comic timing. Like THE MOVES MAKE THE MAN, this is full of memorable scenes and magic moments. A stimulating, thought-provoking tale." —*School Library Journal*

"Brooks gives fresh meaning to love and innocence, commitment and self-realization. And the novel's stunning ending subverts the typical coming-of-age conventions." —ALA *Booklist*

Midnight Hour Encores

Bruce Brooks

Bruce Brooks

A Harper Keypoint Book

Library of Congress Cataloging-in-Publication Data
Brooks, Bruce.
  Midnight hour encores.

  Summary: A sixteen-year-old cellist and musical
prodigy travels cross-country with her father, a
product of the 1960s, to meet her mother, who
abandoned her as a baby.
  [1. Violoncellists—Fiction.   2. Fathers and
daughters—Fiction.   3. Mothers and daughters—Fiction]
I. Title.
PZ7.B7913Mi 1986   [Fic]   86-45035
ISBN 0-06-020709-4
ISBN 0-06-020710-8 (lib. bdg.)
ISBN 0-694-05624-3 (pbk.)

First Harper Keypoint edition, 1988.
Harper Keypoint books are published by
Harper & Row, Publishers, Inc.

ACKNOWLEDGMENTS

Thanks to these friends for always being ready to talk about music in relentless detail: Dan Godfrey, Rich Levy, Kirk Walther, Basil Henderson, Anne Henderson, Dave Pogue, Larry Eckholt, Dan Yoder, JCL, Steve Whealton, Scott Lowe, Doug Hamilton, Doug Gaddy, David Louie, Sarah Metcalf, Don Dixon, Phyllis Dixon, Patrick Hutchins, Chin Chu Hu, Paul Wade, Brent Bingham, Andy Thurlow, Paul Thurlow, Don Kipp, Hank Smith, Glenn Hartranft, Fae Kontje, Geof Kontje, Paul Jones, Michael McFadden, Ken Kekke, Greg Gerew, and dozens of enthusiastic people in record stores. Special thanks to Dan Sperry and Linda Sanders.

For Penelope, Laura, and Gail

Away

1

When I was nine, suddenly all the other girls in the fourth grade were horse freaks. They sat around in class with their hands curved over the tops of their desks so the teacher couldn't see they were drawing deformed stallions on the backs of their spelling tests and lettering sweaty names like STORMY and FLICKER on their notebooks. It would have been bad enough if they were writing boys' names; boys I could have understood. I knew that someday we were all supposed to get into boys. But horses? Nobody I knew had grown up to marry a horse.

I tried to consider liking horses. I even tried to draw a horse myself, after reading about them in a couple of my encyclopedias, but I stopped after the first foot. Horses run on what is really one of their *toes*. It hurt just to think about that.

I decided the reason I wasn't fascinated with horses might be that I had never seen one. I already knew

3

the girls in my class didn't own horses, but at least a few of them had seen them somewhere. So I did what I sometimes do when one of my investigations hits a gap. I asked Taxi.

What would *your* father do if you asked him to show you some horses? He'd say, "Sure, honey," and put down his newspaper or book or *TV Guide* and get you into the car for a twenty-minute drive into the nearby country for a peek at a few of the animals in a local stable. If you lived in Washington, D.C., like us, your dad wouldn't even have to drive twenty minutes, because there are stables in Rock Creek Park right in the middle of town. "There, sweetie pie," your dad would say, waving his arm at a barn or a circled fence inside which a bunch of the things snorted and tiptoed and shivered. "There's some horses."

But my father is a little different. Yes, he put down his book. Yes, he said, "Sure, honey," and got me into the car. Yes, he started driving toward the country. But he didn't stop for almost six hours, and when he did, it wasn't near a stable. It was in the middle of miles of sand dunes, where we cooked dinner over a fire and spent the night in a tent.

When he woke me up before dawn the next morning, he didn't mention horses. He handed me a cup of coffee in the dark and waited while I drank it and ate an apple. Then in complete silence he led me on a stumbly climb across a thousand sand dunes as the sky got less black. We stopped suddenly ten feet from

4

the top of one dune. The sky got lighter at the edges, turning as pink as somebody's gums. Taxi didn't say anything, except to ask if I was cold and offer me his sweater, which I took, also without saying anything.

Then all of a sudden he cocked his head and beckoned me and we climbed with him in the lead. He turned and pushed me up ahead for the last couple of steps. And then *zhing,* I was standing on the very top edge of a straight-fall sixty-foot sand dune, looking across half a mile of pale beach dotted with pools, at the *ocean,* which somehow I didn't expect, even with all that sand. The sun was just untucking itself over the sea off in the distance and the sky was turning pink and yellow. There was a breeze, thick and warm, and I have just forgotten completely about horses because it's pretty nice out there, and who cares about short-haired animals that run on single toes.

Taxi cares, that's who. Taxi cares about anything you ask him about. Put him in motion and you can't just turn him off. He sets things up so well, sometimes you'd think he controls everything, but he really doesn't. He set *this* one up perfectly, though, because just as I am doing my city-girl-marvels-at-nature routine, I suddenly notice that the pounding of the surf is getting louder from a specific direction, the way a secondary theme sneaks into melody from the violas in an orchestra. And when I look in that direction, off to my left, instead of surf I see a sudden wild spray of beautiful monsters from Mars swirl out from behind a dune, gracefully rolling toward me, not snorting or shivering but just *running*, running on the flat beach beneath me, splashing in the edges of the tide and emptying those little pools with a single stroke of a hoof.

Taxi doesn't have to wave his hand and say, "There, sweetie pie. There's some horses." He stands behind me so quietly I don't even know he's there, watching while the herd runs by and runs away, down the empty beach without losing speed, until all I can see is an occasional sparkle in the growing sunlight when one of them smashes a pool. I stare for a few minutes. After a while, Taxi's hand lands on my forearm like a confident little bird, and I turn, and he leads me back to the car. We don't say anything, really. I sleep all the way home.

The next day I went to school, and when I watched the Hilarys and Jennifers write STORMY!!! so urgently, I felt even further away from them than ever. But it wasn't Taxi's fault. I asked him to show me something and he did. And this whole story is just to demonstrate why I only ask him to show me things that seem important. He doesn't fool around, and he doesn't give up, and he doesn't let go even when you would just as soon skip the inquiry midway through.

Three days ago I asked him to show me my mother.

2

Okay, most kids don't need to ask their fathers to show them their mothers at sixteen, right? Mom's out there in the kitchen. Or Mom's out in the car waiting to take you to school. Or Mom's upstairs drunk in bed in front of the soaps, or in the study working on the computer and asking if you think she's only made for looking after you and your friends. She's somewhere. Somewhere there's a Mom.

Not in my house. In my house it's Taxi who's in the kitchen, or Taxi who's warming up the Volvo to run me to my cello lesson, or Taxi who's in the conservatorio drinking too much coffee in front of the stereo playing some moronic sliding-saxophone jazz. Frankly, though, we don't seem to bump into each other all that much—just enough, really. I'm not sure I could have stood having anyone else around to fill in all the little spaces in my life that I move through.

Taxi is never in those spaces. It's like he knows where I need to go and slips out of the way quietly.

At least twice a year for as long as I can remember, and I can remember pretty far back, Taxi has asked me if I want to meet my mom. He asks very politely and casually. No pressure. No disappointment when I say no. And that's what I always say. He says, "Sib, do you want to visit your mother anytime soon? She lives in San Francisco." I think it's nice the way he sticks the city into the proposition, as if that could legitimately matter to me too, because it very well might, since I'm such an incredibly picky person. "No thanks, Taxi," I say, and it's never because she lives in San Francisco.

I have never asked much about her, though I'm sure he would tell me. Maybe I just never had the time to start an investigation or go through one of Taxi's big numbers if I did ask him. Maybe I thought it wouldn't teach me anything I had much use for, since she wasn't around anyway. Maybe I'm careful with my curiosity. The fact is, I've lived with Taxi for sixteen years and I haven't gotten all that curious about *him* yet. We get along. The spaces are open. That's always been enough, nice and easy, no surprises.

Until last Saturday. But last Saturday I shocked Taxi, really buzzed his wires. I set him up pretty carefully, but he still wasn't prepared.

Music has taught me that people are suckers for

8

foreshadowing. Give them a little lyrical fragment in the woodwinds behind the big theme in the first movement, then bring it back through the double basses in the second, and by the time the whole melody jumps out in the third movement, they feel like they wrote it themselves and have been waiting to hear it for years.

So I made a couple of special little shifts in our Saturday routine to give Taxi his lyrical fragment. First of all, I got up early. I usually sleep until almost ten on Saturdays because my improvisatory chamber ensemble on Friday nights goes until after eleven, and when I get home I usually want to notate some of the things we discovered during the evening. This duty falls to me since nobody else in the group is the least bit interested in working that hard. Roger, who is a perfect jazz jerk but the only trumpet we could find, even makes fun of me for notating, saying things like "Hey, when you improvise, it's here and then it's gone. That's the nature of The Music. Like life and death, honey." Roger is eighteen and he feels forty-four. I'm sixteen and I feel sixteen, and that means there's a lot I want to learn, even if I already know more than Roger.

Taxi once asked if perhaps we wouldn't "be more comfortable" setting an earlier time to break up. I reminded him we *had* set an earlier time when we started the group—we said we would finish every night by 9:30. We had to do this to make Effie Pee-

9

ters's mom let delicate Effs come and play her fragile oboe with us. But improvisation is different from written music. If you're going to play the Mendelssohn Octet, you know how long that takes and you know whether you'll catch the 9:16 or the 9:46 bus home. But when you're making everything up on the spot, you never know where it's going to finish. Sometimes the best stuff comes from one person just when everybody else thinks we're about to crap out. And then of course there's Roger, who stretches his solos out three times as long as they deserve just so he can play endless whole notes and show off his circular breathing.

When I sleep late, Tax gets the early part of the morning all to himself. I know he enjoys it. He drinks coffee and listens to jazz records and rereads the printouts of his newsletter even though it went to the printer the night before and it's too late to correct anything. He does all this in the conservatorio, which is what I call the front room because that's where all the music is: the stereo, the records, the piano, and my two fool-around instruments, a viola de gamba and a double bass. I keep my cello in my bedroom.

So he was surprised to see me walk into the conservatorio early Saturday morning. He was standing by a front window, where he had just raised the shade, letting a glassy sheet of sunlight creep over the sill from the east. When he saw me, he gaped and said, "Wow."

10

His coffee mug was sitting on a little end table beside the chair in the middle of the room at the optimum conjunction point for the sound from our speakers. The newsletter printout, sure enough, was there beside the mug, and, sure enough again, there was some drippy solo piano jazz on the turntable.

I picked up the mug and slurped a gulp that diminished its contents by about a third. Taxi makes strong coffee but he drinks it with milk. I like it black.

"Want a cup of coffee?" he said, trying to be a little ironic but without teeth; he's not mean enough to be really ironic. I am, but I don't get it from him.

"No thanks," I said. "What's this music?"

"Roland Hanna, playing piano."

"So *that's* a piano," I said. What a jerk I can be sometimes. But it was so early and the light was making me squint.

Taxi, of course, did not look hurt. He never does. He leaves it up to me to recognize when I've been nasty, all on my own. Instead he said: "Is it too bright for you, maybe?"

"I thought you'd never ask."

He lowered the shade halfway. I grunted. Then, to be nice, I said, "Actually this tune is kind of interesting. The harmonies remind me of part of the Debussy Quartet."

He beamed. "Hey, I was thinking it sounded like Ravel!" He looked at me hopefully. "That's pretty close, isn't it? I mean . . ."

"Close enough," I said. "Because the Ravel Quartet reminds some people of the Debussy Quartet, so there you are."

He looked like he could be pretty happy about it, but of course he had to ask: "Some people? Does it remind *you*?"

"Not really."

He sank a little and nodded. Then he watched while I slurped the rest of his coffee and walked to the other front window. I looked through the blinds. There was a red bread truck pulling away from the bakery across the street, and its engine was whining in the same pitch as the first notes of the third Bach Cello Suite. *This is the moment,* I thought.

"Hey," I said. "I want to meet my mom now."

I looked to the side and met his face full on before he could hide the stammer in his eyes. He opened his mouth to speak, but I cut him off.

"Not, you know, right now. Not today," I said. "But in, oh, ten days or so." It sounded a little too nonchalant; I was trying too hard. He was studying me.

Taxi studying me was worse than Taxi asking me what I was up to, straight out. We have kind of a tradition of keeping secrets from each other—respecting privacy and not meddling. When one of us does ask the other for something that obviously isn't being offered, it's fair for the person with the secret to answer with a snap or even a fib. That takes care of that question for good. But in our unwritten rules, it's

12

okay to study and ponder and suppose, instead of asking, as long as you don't get too direct. With Taxi, that's dangerous for your secret—he's pretty smart.

"Whenever you say," he said cautiously but sincerely.

"I say in a week. Or thereabouts." He nodded slightly, knowing I hadn't finished. I sighed and decided to let him have it blunt and mysterious: "I just want to be in San Francisco on the twelfth of July."

"Oh. The twelfth. Sure. No problem. Plenty of time." He looked as if having time to get there had been all he was concerned about. I knew it wasn't. Well, he could guess all he wanted—I knew he wouldn't get this secret until I was ready to tell him.

"All right," I said brightly, "that's settled. Now we can get an early start on the vadgers, can't we?"

He blinked, refocusing. Since I was a tiny kid, we've hit the streets together on Saturdays and explored the town from late morning to late afternoon. City adventures, which I called *silly vadgers* before I could talk decently. The name stuck. "Yes, right," he said, nodding belatedly. "Vadgers."

"Vadgers to be had, on the ribbon of moonlight." My favorite poem used to be "The Highwayman," by Alfred Noyes; I memorized it in the third grade and recited it for a PTA meeting. The ribbon of moonlight was what Noyes called the road.

"Looping the purple moor," said Taxi. Which is where Noyes put his road.

13

"Tlot-tlot." Which is what the Highwayman's horse sounded like on the bricks of the ribbon of moonlight looping the purple moor, and so on—Taxi and I know the whole poem, and we could have gone on swapping quotes for minutes, but Taxi pulled out early, which he *never* does.

"Okay," he said. "Breakfast first?"

"Breakfast out," I said, turning and hitting the reject button and thus jerking Roland out of his augmented fifths in the middle of what I recognized too late as the first theme of Taxi's favorite song, "Chelsea Bridge." I also realized a little too late that the harmonies old Roland was using did sound like Ravel.

"Hey," I said. "That really *did* sound like Ravel. Not Debussy. You were right."

"Sure," he said. But he didn't believe me.

3

I'm finding out that writing's not too bad, but I still don't read much; if you've made it this far you're a better reader than I am. The reason I don't read is that books are so damn locked up. To me, the little black words on a page are stiffer than steel forks, more closed than the stones in the Great Wall of China.

When I'm reading, I don't like the idea that I can't stop the yarn spinning for a second to ask a couple of questions, to clear up a point or two. Sometimes I'd like to reach into the text and switch a couple of things around, throw the spotlight on a character I think should be getting more play, or shove the narrator off in a new direction. But I can't—it's all so *finished*. You can have the story only one way, the author's way. That's tyranny, isn't it?

Well, here I am, writing a story, and it looks like you're going to get it pretty much my way. Suddenly I'm a tyrant too. Now I understand the problem. I can't see how to let you in, to let you change things around.

Music is different. Music is written down, but it's not stony and stiff—in fact, the guy who wrote it *wants* me to fool around with it a bit, to poke into the special places I discover, to set my own pace, to tell the story with my own accents. So it's lucky I'm a musician, not a writer.

I play the cello. I'm very good. If it means anything to you, the circle of international music critics puts me about third or fourth in the world right now. Some say Yo-Yo Ma is better, some add Janos Starker to him above me, some say I beat the wings off both of those guys but take a whipping from Lynn Harrell or old Tortelier, who is probably my favorite player though I like things about all of them: Yo-Yo's grace, Starker's will, Harrell's sense of design. But of course Tortelier's not truly my favorite player. *I'm* my favorite player; otherwise I'd play like Tortelier, which I don't. I do have trouble getting away from a few of his tricks when I do the Elgar Concerto, though.

It may seem pretty cocky of the critics to think they have the authority to rank musicians as if they were football teams. Actually, they've got a little something to go on besides their own impressions: international competitions in which musicians play for prizes. I went to the Brussels competition three years ago and won; I went to the Prague competition two years ago and won; I went to the Rome competition

16

last year and won that too. It's bad manners to win any more, and even by last year I'd kind of outgrown the circuit—the contests are for people who still have to prove they're hot stuff.

Earlier this summer I took exams to place out of my senior year of high school so I can go next fall to Juilliard (where I'll probably stick for only a semester—I've really kind of outgrown Juilliard too, but it's a Taxi-approved way of setting up in New York).

There are some good things about my school, but I won't miss it. I like a few of my teachers and a few kids. There are some kids who are as clever at playing School as I am at playing the cello. They find their way around all sides of each teacher and each subject with the same kind of cunning and arrogance (to use two of the words that keep following me around in my reviews) that I use to learn a new piece under a big conductor. I like watching them operate, but they don't really teach me anything. Not like the journalists.

One of the things that happen on a concert tour or at an international competition, when you're a teen prodigy everybody looks at as either a freak or a threat, is that you get to pass the time of day—of a lot of days—with reporters.

They're a pretty interesting bunch. More interesting than all the high-strung kid musicians with ulcers at twelve and alcoholism at fourteen. The reporters have ulcers and alcoholism too, but they've done more things to deserve them. The kids are a mess because people treat them like horses: train them to run beautifully on one toe.

The reporters really ticked me off at first, the way they let the usual human manner go, grabbing me and asking the same question twenty times until I gave the answer they were expecting before they even met me, trying to charm me with every kind of false face and voice, hollering at me when I took a breath during an "interview" six minutes after coming offstage from playing the Dvořák Concerto with the Berlin Philharmonic.

But I started hanging out with some of them during off times, and I found out they were smarter than they acted. They act like they know nothing just so you'll try to tell them everything; then they pick out what sounds spicy and print it. Not a bad ploy, really.

The thing they are smartest about is language. Whatever I know about writing I learned from them. You'd never think it from reading their stories, which are either too melodramatic or too cold—but that's the nature of newspapers, not reporters.

You wouldn't believe the elegance those crunchy old types can come out with over a game of cards or a couple of glasses of whiskey. They can talk for hours and never lose the flow of grace and wit and rhythm. Especially the ones from Ireland and England and a few from the U.S. The ones from France and Italy and Brazil get to be more wild in print, so they don't have to save up for talk the way the English-language guys do.

I ride on the press bus between concerts sometimes, playing poker all the way from Munich to Geneva or talking soccer from Vienna to Milan. I've learned a lot of words and a lot of weird sentence structures and a lot of dirty jokes.

I suppose the reporters are the best friends I've got, though they're the kind of friends you have only when you're with them. I have no friends at school, really—I'm too picky. I have Taxi. But he's my father, so it's probably not fair to count him as a friend.

I suppose he really fits somewhere in between. That's probably why I don't call him by a daughter-only name like "Dad," but I don't call him by his real name either. His real name is Cabot, Cabot Spooner. Once when I was small I heard somebody call him "Cab." I got the idea that calling him "Taxi" was a great joke after that. Naturally I repeated the joke a million times, and we both got used to it.

Taxi calls me Sib. For Sibilance. That's been my name since I was eight: Sibilance T. Spooner, and the T. doesn't stand for anything.

The name I had before that was Esalen Starness Blue. How do you like that? I didn't. Esalen is a place, in California I think, where people go to hang around with gurus and shout confessions at each other. Starness you can figure out: "the quality of being of or like a star." Whether this was meant to be a star in the sky or a star of stage and screen I don't know. And to tell the truth, one thing I dislike about being a famous musician now is that once in a while somebody calls me a "star," and I can't help wishing this didn't make my original name seem legitimate. I mean, Starness. What a crappy idea. As for Blue, it means the color. Naturally, until I changed my name I always thought I looked awful in blue and never wore it. I *still* never wear light blue—it makes me feel like I'm floating in a swimming pool.

When I was eight, one night Taxi said just out of nowhere, "You know, you can change your name if you like." I had stopped bitching about it years before, since I was old enough to hate it; the less anybody said about it, the better. So why he thought to say that night I didn't know, but he was weirdly on target because just that week I had been thrashing around inside that name like it was a wet wool coat worn inside out against my skin—but in silence, always in silence.

I asked him then what Esalen meant; I never had dared to before, afraid that finding out the meaning of the one word in my name that I didn't know would be like plugging up my only escape hole.

He told me. I asked how I got it. He hesitated for a second, then told me that my name was given to me by my mother. It was the only time except for his regular formal invitations that we had ever mentioned my mother. She popped up into the conversation like a cardboard parrot in a kids' book, and I asked why she had given me that name; he said that she had spent a week at Esalen once, and it had been very important to her. I asked why she had added on Starness. He said he thought stars were very significant symbols of all kinds of terrific stuff. Why blue? He said it was supposed to be my favorite color. Supposed to be, according to whom? He sighed; according to either my astrological alignment or the birth mythology of the Hopi Indians, he didn't remember which.

I told him I would be happy to change my name.

It took me a few days. I won't bore you with the names I considered, but I will say that I ranged pretty far away from the likes of Esalen Starness Blue. I even considered Mary; that was about as far away as you could get. But none of the names I considered fit.

Then I realized that not *all* of my old name felt like an artificial leg. There was one syllable that had come to suit me okay, because it was all Taxi had ever called me: *Ess.* It was how I introduced myself to kids at school, too—"Call me Ess."

I decided to use Ess as the basis for my new name. I didn't pick it as the name itself because it had some problems—it tended to invite nicknames like "Esso" and "Ess-Oh-Bee." But as a start it was fine.

I fixed on the letter S. But all the S names I tried sounded stupid—Samantha, Stephanie, all those jerky things girls name themselves when they're pretending to be princesses in magic kingdoms. I would rather have been named Stormy.

Then one day I came across the word "sibilant" in a crossword puzzle. I looked up what it meant, and suddenly I knew I was close. Here was a word that not only started with S, but was the very definition of S! I tried "Sibilant" for a few days but it wouldn't stick—it sounded too much like a Knight of the Round Table.

Looking through one dictionary, I found the noun form, and then I really had it. *Sibilance*. I loved it. I still do. I think it's the perfect name for me in every way, right through all of its meanings, right through all of its sounds, right through the way it looks on paper. *Sibilance T. Spooner*.

The Spooner part is for Taxi. What the heck, better than Blue. And the T.—that was an afterthought. It doesn't stand for anything. Taxi actually suggested it; he said "Sibilance Spooner" sounded like the author of a slushy romance novel. I asked what kind of books "Sibilance *T*. Spooner" would write, and he thought for a while and said, "Something like an analytical contemporary history book." I said I would never write one of those either.

4

A few mornings later I'm packed and ready to roll. My clothes are squeezed with my usual fanatical neatness into two of the flashy black American Tourister suitcases Taxi bought me before the Prague competition. The bags are stacked just enough out of the way at the top of the stairs, ready for Taxi to carry down when he's ready and stick in the Volvo.

But when I come home on the bus from my last lesson before the trip, I find instead, at the top of the stairs, one lumpy green canvas thing, a log with a zipper in it. It smells like fish. I unzip it six inches, enough to see the edges of a couple of my skirts inside.

I find Taxi in the front room, reading the classified ads.

"What's that green thing with my clothes in it?"

He snaps the paper down a little nervously. "Oh. Hi."

"Hi. What's with the green bag that smells like whale meat? Where's my luggage?"

He thinks for a second and says: "It's a duffel bag. I've got one too. I looked all over the city for them. Finally found them at the Sunny's Surplus store on 7th Street."

"Fantastic. Congratulations. And did Sunny swap you even up for my Touristers?"

He smiles. "Oh no. Your bags are still here. In your closet."

"Why the switch?"

He wrinkles his forehead. "Well . . . the luggage just isn't . . . appropriate."

I snort. It's something I do sometimes. It's not so ugly because I've gotten pretty crisp about it. "Appropriate? What are we going to do—invade Cuba?"

He laughs. "There is a certain irony in the use of army things . . . but that's for later. Let's just say that the suitcases are too . . . formal. In their structure, I mean. Even if they weren't black. But the fact that they *are* black really disqualifies them. Nobody back then would have been caught dead in anything black." He shakes his head, and before I can ask what "back then" refers to, he goes on. "Literally caught dead. I remember Billy Gallagher's funeral. He died playing an electric bass solo in the rain at a local rock festival. His pallbearers painted his black coffin with Day-Glo paisleys and wrote 'Bottoms Up' in airplane

24

glue they sprinkled with glitter. The bass player back then was called the 'bottom' of the band, see . . .''

"Back when, Taxi? What are we talking about here?"

He snaps out of it. "Back a while ago," he says evasively.

Great. That explains everything.

"And anyway," he goes on, "your clothes need a few wrinkles. A healthy informality that refers back to . . . to the days before the current run of narcissism."

The last thing in the world I want when I meet my mother for the first time in my life is a wrinkle-free dress. Thank goodness Taxi is on the lookout against narcissism. But the army bags are only the beginning. Next it seems, we need a bus.

Not just any bus. We have to find a 1965, '66, or '67 Volkswagen, and not just any old '65, '66, or '67 VW bus either. The first two that we see—using the want ads—are missing something: an insignia on the front, special curved windows in the back, or some such nifty little detail. One bus Taxi dismisses because it doesn't smell right. "All buses from this era should smell like clean oil, day-old homemade yeast bread, and wool lint," he explains, not just to me, but also to the poor guy who is trying to unload this bus, which his brother left behind four years ago when he joined the Peace Corps and went to Botswana. Taxi demonstrates how one uses the nose to sniff. "You

25

can smell for yourself that it isn't right." The guy drops his price to $200. Taxi sighs and says he's sorry. We leave. The next bus we avoid because it has been repainted reddish orange and Taxi senses it's "not at peace."

"I thought 'at peace' was a synonym for 'dead,' " I say as we drive off.

"Good buses never die," he says.

Finally in Chevy Chase we find a blue bus sitting in the dark outside a big house with an open, well-lit garage full of silver BMWs. Taxi grunts as we eye the familiar box perched on those funny thin wheels that slant outward at the top. "Out to pasture," he says.

A guy in his mid-thirties lets us into the house. He's going bald but his hair's pretty long on the sides. He's got on one of those sweat suits that cost big bucks nowadays because the ribbed circles around the neck and wrists have been ripped out, and silver running shoes. The first thing he tells us is his wife's out doing Nautilus or he'd offer us some tea. He can't boil water by himself?

Taxi asks him about the bus, but the guy really talks about the BMWs instead. "Oh, she's been a loyal old getabout," he says, "but obsolete now that we have the Bavarian stallions, obviously." Horses again.

We go out through the garage. The guy walks warily past the silver cars, pulling in his stomach and

26

turning sideways so that he doesn't touch any steel. Taxi looks a little nervous, expectant.

We come up behind the poor old thing in the dark. The guy says, "Wait, I can turn the spot on," but Taxi says, "No, really, don't bother. I want to see it this way first." This way? In the dark?

But Taxi seems to know what he's doing. He's flitting all over the bus like an ant looking for a crack in a walnut, and from the way he mutters to himself I can see he's finding things okay without light. He sticks his fingers in small places between pipes underneath and smells them, he pulls on parts of the rear and side panels, he presses all ten fingertips on each window and releases abruptly. He whips out this little gadget and does what he tells us is a compression test on the engine. He jacks the bus up and fools around with the rods and things surrounding the wheels. Finally, nodding to himself and wiping his hands on a rag he must have brought in a pocket, he says, "I'd like to spend a few minutes inside, if that's okay."

While he's in there doing some secret test, the owner tells me a long story about how he bought the bus as a college student in North Carolina and planned to pay for it by running factory-outlet cigarettes up to Philadelphia. The first time he packed the bus full of N.C. smokes and headed north for glory, he got busted. Seems the turn signals on the bus started flashing in crazy alternating patterns until a Jersey Turnpike cop pulled him over, roughed him up, and confiscated the cigarettes. "The funny thing was, I never touched the turn signals—they went on by themselves." He laughed uneasily. "I guess this bus has an honest soul."

Then Taxi is back, looking enlightened. "Good bus," he says to the guy, very soberly.

"Yeah," he says. "A classic, really. An antique!"

"Not for another year," Taxi says mildly.

"Runs great!"

Taxi nods neutrally. "Needs a ring job, and the solenoid is shot."

The guy can't think what to bring up next. I offer brightly: "The turn signals work all by themselves!"

The guy grimaces. We get it for two hundred bucks.

I follow Taxi home in the Volvo. I don't have a license; I never seem to be around long enough to memorize how many car lengths you're supposed to follow behind a dogcatcher truck with its light flashing and all that stuff, but I drove a Volvo bus in the Alps once, so I guess I can drive a station wagon through Chevy Chase.

When we get home I ask Taxi what he did inside the bus for fifteen minutes. All he tells me is that this bus has a great soul.

"Why don't you call it 'her' like the other guy who believed in its soul?"

"For the same reason I never call you 'it,'" he says, giving me a kiss on the forehead.

This is probably the right moment to ask Taxi what he's up to. What the duffel bags and the bus and the talk of "back then" mean. But I'm tired, and he's exhilarated. Besides, I'd probably do better to let him keep his secret for now. That way he won't look too hard for mine.

5

In 1947, exactly thirty-six years before I won the same two prizes at the same age, a thirteen-year-old Russian boy known only by the single name Dzyga swept through the revered Brussels concours and beat the crap out of everybody by winning both the cello first prize and the festival's grand prize as the top instru- mentalist among all competitors on all instruments.

My repeat of the sweep called up the Dzyga specter in the mind of only one person, an old Castilian re- porter covering the competition. He was at least sev- enty, very leathery but graceful and courtly, the way old Spaniards from that district are. Casals, of course, was Castilian. This reporter was the first person to mention Dzyga to me, though he didn't even use the name: *You have taken your place alongside a ghost,* he said.

I got curious. But all the old man could tell me was the boy's name, the fact that he won at Brussels and won the next year at Rome, and the fact that he had never been heard of since.

The people I talked to take the disappearance as normal Soviet behavior—either Dzyga's or the government's. "The best Russian musicians hate being the best Soviet musicians," said one orchestra violinist who thinks Dzyga dropped out on his own. "If they're bright enough to play music well, they're bright enough to hate doing it for the greater glory of their lead-brained regime." The other people, the ones who think Dzyga was suppressed, shrug and say, "It's Russia."

In Prague a couple of years ago I talked to a few of the Russian string players; last year I saw a couple of the same ones, but they steered clear of me. Not just because they knew I was going to clean their clocks in the competition, either: They were *instructed* to keep away, one of them told me in secret, because of something I had said in Prague. The only thing I had said to them in Prague was "Hey, you guys ever hear of a cellist named Dzyga?"

Several months later I came across a nameless mention of the prodigy in a footnote of an old music criticism book I was thumbing through in a used-record store. I wrote to the author of the book, care of his publisher; I got a letter back from a conglomerate that had bought out the publisher and dissolved it, telling me the author had been dead for twenty years. I went to a couple of libraries and music bookstores, and searched everywhere for mention of Dzyga. Nothing.

I went to the Library of Congress next, and spent a full week of eleven-hour days sifting through catalogued and unidentified concert or recital recordings from the late Forties and early Fifties, especially those made in Europe. Lots of them were awful, but I listened long enough to decide Dzyga could not be there.

Then one day I found him.

It was an old wire recording of a pickup orchestra of the kids who had competed in the 1948 Rome competition, playing the Haydn Cello Concerto, obviously sight-reading. The orchestra was horrible, full of broken hearts and fatigue; the conductor—probably also a kid—was out of control; the sound quality of the recording was harmful to the tiny bones. But the cellist . . .

When I heard his first notes, I stopped being a musician. I stopped being smart. I stopped being tall and lonesome. My outlines and edges faded away and spread out everywhere like light and water. There was nothing but this sound worth my notice.

Then his playing *pulled* on me, deep in my chest and legs and eyes, and I felt as if I knew nothing, as if my body was just being made. I sat there and shivered and blinked, like a baby hearing a voice it loves before it knows words.

His playing was so good it was almost wild, as if the boundaries of music were stretching. Yet finally, he was in perfect control of himself. Lots of people have very precise pitch. Dzyga's was better—almost original. He found the notes by some secret path. His sense of rhythm was free, almost improvisatory, emphasized by what I first thought was a hotdog thing: an almost complete lack of vibrato. *Hey, I don't need to feel my way into the notes, folks—I nail them and hold them.* The sustained notes were kind of eerie.

The most difficult thing to pin down in terms of technique was a way he had of making it clear that every note was coming out of silence, a surprise. Not one tumbled logically from the note before it. He went back to the nothingness before every stroke. The only thing I've ever heard remotely like it was a tape my improv pal Roger once played me, of a pianist named Thelonious Monk. But Thelonious Monk was playing quirky little ditties he wrote; Dzyga was playing *Haydn*!

He had been a ghost; now he was a sound, an amazing sound. I listened to him one more time and put him back into the vaults of the library. He wasn't safely tucked away anymore. He was out there, somewhere, in the open. And nobody knew it but me.

6

I find Taxi out in the bus the next morning, tinkering with a bunch of wires around the steering column. He grins and makes a show of touching two of the wires together to make the geeky little horn beep three syllables that I guess stand for GOOD MORN ING. After knowing quite a few Germans in Europe I can't believe they would stick a horn like that on something that was supposed to be a bus. It sounds like an insect clearing its throat. An insect that doesn't bite.

"Going on any errands today to pick up stuff for the trip?" I ask.

"Need something?"

"I need a stool," I say. He nods and waits. "A stool that seats me between eighteen and twenty-one inches off the . . . floor. One that will stay stable while we drive."

He wrinkles his forehead. "You want a stool for the back of the bus?"

I nod.

"Can I ask what for?"

I swallow. "Practice."

He gapes at me. "To practice your cello?"

"No, to practice sitting like a lady. I figure I need all four days to learn to keep my knees together in all the various seated positions."

"Sib, you . . . I know you want to make a good impression at Juilliard, but that's not until September! You can take a week or two off, can't you? I mean—that beautiful instrument rattling around in here . . ." He shudders.

"That's why I need a good stool," I say, and turn back for the house.

"At least take Grunt Martha instead of the Bianchi," he calls after me.

"It'll be the Bianchi," I tell him over my shoulder. "It's the one that has the good soul."

Later, while listening to a new CBS recording of Ma doing the Bach suites—and coming nowhere near Starker's old Mercury version—I think I hear a clattering in the distance that couldn't possibly be blamed on CBS's crummy digital technology. I flip the record off and the clatter is still there. Taxi is rummaging in the wood in his workshop out behind the apartment.

I decide to spy.

It's easy enough. His workbench is sort of under a high window, by which an old exterior staircase passes.

The staircase isn't in great shape, and it ends abruptly in space, but I've crouched on it many times to watch Taxi make things. The sanding, planing, and sawing used to soothe me in some way when I was a kid. I never knew what Taxi was making; I just watched birdhouses or Ping-Pong paddles or small bookshelves take shape. They all started from a plain piece of wood that could have been the same one every time. In fact I used to half pretend, half believe there was one plank from which Tax had to cut the first piece of anything he wanted to build. If he didn't cut it from the old Master Plank, it wouldn't work. If he did, and he finished building the thing, the Master Plank would be whole again the next time he needed a piece.

As I climb the stairs I notice it's been a long time, probably since I took up the cello—my legs reach different steps now, and I have to take more care to be quiet. It's awful what a spider I've become. I've always been tall, the tallest girl in my grade since second grade, when Aggie Dalrymple moved back to Scotland. She was a freak, though, five feet tall at seven; she's probably in a circus by now.

At least I have thick hair. It gives me some substance, even though I sometimes get tired of it. If I get caught up in a piece I'm playing and let my head bob and hug too close to the head of my cello, it gets wrapped around my tuning pegs, and I want to cut it short. But then I worry I'd look like a pinhead. I have big feet, and I don't want my head to look smaller than they do.

But I'm not a nut about my appearance. It could be a lot worse, believe me, when you consider that forty to sixty nights of the year I'm on stage in front of strangers, hugging an awkward-looking wooden thing between my knees and scratching it with a long rod. I met a few appearance nuts early in my playing days, and two of them had breakdowns and quit in their late teens. I saw one of them flip out before a crummy small concert because she thought her shoes let the *backs* of the heels of her stockings show. And she wasn't even a soloist that night. She ended up going to a local junior college and studying cosmetology. It's not just girls, by the way—I've known some boys who cracked up, mostly over thinking they were getting bald spots.

I fold myself onto the steps next to the window and look down at Taxi. He's cutting his first piece, from the Master Plank. I can feel the sawing all the way up through my bones. It's still soothing.

So is looking at Taxi again. I realize I haven't really watched him in a long time. I guess I got to know him so well—what he looked like, how he moved— that I stopped noticing, taking it for granted that, for example, when he was cooking an omelette he looked like Taxi Cooking an Omelette. Beneath me now is Taxi Making Something from Wood. But I don't take it for granted today. I watch.

He's pretty nice looking, actually. Thick brownish hair that I usually wish he'd gotten cut the week before we go anywhere together. Blue eyes that seem to throw a little light when they zip to something. Lots of lines on his face, which I suppose you'd call wrinkles, technically; I can't bring myself to call them wrinkles, because they come and go so quickly and always leave him looking young. I've seen him asleep a lot, and when he's asleep there isn't a line anywhere. But when he smiles a certain way or frets, his face is full of them.

He's thin, and he moves very neatly. A lot of spring, and I've seen him run incredibly fast. You'd think he'd be good at some sport or other, but you couldn't pin him down to any single one: too short for some, not enough bulk for others, a lack of attack and roughness. I see a lot of jocks in school, and he's unlike any of them. Once at a camping long weekend for parents and kids I saw him beat the crap out of every father he played in tennis, basketball, pool, and horseshoes—but all the other guys felt great about it because they all *looked* as if they'd pound him nine times out of ten and it was a minor fluke. But he just kept laughing and being nice and beating people. The only thing I saw him lose all weekend was a swimming race. When they gave awards at the camp banquet he was chosen Best Athlete, and he laughed about it and so did everyone else, and they all clapped hard because they liked him. He laughed too. Ho ho, ha ha, see what you can do with this backhand up the line, that set's mine 6–1, I believe. Next.

He wears pretty dull clothes most of the time, khaki pants and pullover sweaters and sneakers. A few days a week he has to dress nicely to go dig up obscure inside stories on Capitol Hill, and he always surprises me by how dapper he looks in a suit. I can't get used to it—suits just aren't Taxi, no matter how good he looks. It's always as if he's making believe.

But right now, on the bench in the workshop, he's making something real enough. Cutting small pieces of different woods and fitting them into weird things, brushing a clear liquid carefully onto some of the poles and platters he has sawed and sanded, and setting them off to the side to dry. His hands change between tasks— when he saws they're compact and inflexible; when he sands they flutter and curve like bird wings. His face, though, is always the same, focused on what he's doing like a fixed lens.

After an hour or so my knees and ankles demand a change, and I'm glad to give it to them. A few years ago I would have stuck it out to the end. I guess I'm out of practice.

7

One of the things I inherit directly from Taxi's chromosomes is his talent for getting the small things right. He's a great detail man.

Last night while I slept, Taxi made the whole house into a stage set for the play *Sib and Taxi Are at Home as Usual*. He rigged up a coordinated system that simulates our presence in the house, with lights and radios in three rooms and even a 90-minute tape of me practicing, which will play twice a day in the conservatorio. Before breakfast I got a tour and demonstration. Taxi was excited. I was amazed.

It's pretty hard to amaze me on the morning of a trip. Usually it's pretty hard to get me to do more than grunt mean grunts and squint mean squints. When I wake up to hit the road, there's extra crust in my eyes and extra stiffness in my back and extra lemon juice in my outlook on life. I like seeing different places, but I hate getting there. On concert tours I spend more time getting there than I do in the places, which makes me furious—it seems so inside out.

But I feel different this morning, pretty fresh and curious. Naturally. There are unknowns out there in San Francisco. This is going to be one of my most surprising journeys, I figure—and I like surprises.

After breakfast, as we're leaving, I get a pretty big one.

We step outside. Taxi slams the door behind us, and I feel a sudden twinge of farewell for the digs. But I don't hang back, because this is one of those D.C. summer days when the sunlight feels like pound cake—yellow and heavy. So I flee toward the bus, wrench open the side door, and scramble in.

The whole back part of the bus has been cleared, except for the cello and some junk along the sides. In the middle of the metal floor lies a thick rubber mat about four feet square. I try to peel up one corner, and find that it's been glued with something fierce; it won't budge even a little. And on top of the mat is this piece of crazy furniture.

It's definitely a stool, made out of some pinkish wood, but you could be excused for thinking it's a model of a prehistoric bug. It has six legs, for one thing, but its most creepy feature is what's on the end of each leg: a claw. A genuine set of curved iron talons squeezing into the mat. I'm trying to remember where I've seen them before, when Taxi's voice comes over my shoulder from the front seat:

41

"Piano stools," he says. "The old round kind that screw up or down."

"Right! But they usually have glass balls in the claws."

"I had to chip the balls out with a hammer and cold chisel. Toughest glass I ever came across. Amazing."

And suddenly I recognize a couple of the struts and metal hinges from my spying session yesterday: yes, this is what Taxi was building in his workshop. Designed it, too. "Fast work, Tax." He must have been up all night. "Thanks."

"Hope it works okay."

"I'll try it out when we're moving."

"This rubber should be pretty good at keeping the axle noise out," he says, starting the engine as if to demonstrate. "And I caulked the windows."

"We'll smother. We're going to drive across deserts, aren't we?"

He taps a chrome box under the dashboard. "This is the air-conditioning unit from a 1979 Cadillac Coupe de Ville. Picked it up at a junkyard yesterday and wired it in." He grins. "The guy at the junkyard assured me that in ten minutes it would chill a warm beer on the *roof* at high noon."

"I'll let you know when I want a beer." I sit on the stool with my feet in position to play, and spread my arms as well; the bus spins and bobs up Wisconsin Avenue, and I am relatively immobile. The rubber serves as a kind of buffer that keeps me from being jolted by turns and bumps, and the chair holds to the rubber like an ambitious owl I once saw trying to pick up a small deer for perhaps a quarter mile of open field, the deer running frantically, the owl beating its wings and lunging upward; it finally gave up and just rode the deer for a while to catch its breath, then let go and lifted off.

Suddenly Taxi says "Oh!" and smacks the steering wheel. For him, that's a tantrum.

"What? Somebody cut you off? Or did you leave the bathwater running?"

"Worse," he says grimly. "I forgot something."

"What could it be?" I say, poking around at the junk in the back. "We've got a shredded tent, we've got a greasy cardboard box full of old tools and bus parts, we've got two duffel bags already hard at work getting rid of our narcissism, and we're carting the whole pile across the country in a twenty-year-old piece of machinery designed and manufactured before the Germans recovered their self-respect. What more could we want?"

"The music," he says, almost to himself. "How could I forget the music? Without the music it's nothing."

"What music? And what's nothing without it?"

He doesn't answer. Instead he cuts across two lanes and whips into a small shopping center. "Be right back," he says, and jumps out of the bus and into a phone booth. He flips through the Yellow Pages and in twenty seconds he's back, looking relieved. "There's a place in Bethesda," he says.

"Fabulous, Tax. What kind of a place?"

"Used-record store."

Now we're talking. I can get enthusiastic about hitting a used-record store anytime. But I already know every used-record store in the area, and I've never heard of one in Bethesda.

"What are we going to do with records? Did you pick up the turntable from an '84 El Dorado at the same junkyard?"

"We'll have to go home for a couple of hours while I make some tapes. Then we'll buy a tape player."

The thought of recorded music gives me a second's yearning for the Volvo, with the $1100 stereo I so generously installed with part of my winnings from the Rome competition. My tapes of old RCA Chicago Symphony records would be even better inside this metal crate—Reiner's trumpets at the beginning of *The Pines of Rome* would really rattle some screws.

Before I know it we're pulling up in front of a shabby old fake Tudor house off Wisconsin Avenue.

"Perfect," Taxi says, hopping out and smiling up at the cracking stucco and a front lawn as bumpy and brown and splotchy as a meat loaf I once tried to make for dinner.

He looks at me and cocks a sly little grin. "Okay," he says. "Let's call this Lesson One."

"In what class?"

"Archaeology," he says over his shoulder as he practically trots toward the front door. I can hear some music dribbling out. "The exhumation of an Age."

"Very Low Baroque?" I ask, looking up at a busted window behind which a bare light bulb shines in the second-story front room.

He's already scooting up the stairwell, and only part of his answer makes it back down—I catch what sounds like the word "Aquarius." Means nothing to me. I'm a Capricorn.

I actually bump into Taxi at the top of the stairs, just standing there grinning and taking in the atmosphere. He sighs happily and turns to me. "Dig it," he says. "We are there."

"Where?" I say, but now he has darted toward the record bins. I look around. As far as I can see, it's just another used-record store, with a few weird features. Incense in the air. An overstuffed purple sofa with one front leg replaced by two chipped bricks. The obligatory two loudspeakers standing too far apart and pointing in different directions, but with the nice touch of not matching—one's a KLH and one's an Acoustic Research. A whitish cat is asleep on top of the AR, and a front paw dangles down in front of the tweeter.

Behind the counter is the standard used-record-store clerk, with the required flannel plaid shirt and heavy eyebrows, except this guy has a ponytail lying in a hunk on his slouched back like a dead rabbit, and a

45

moustache that grows straight down past his lower lip and covers his mouth entirely. Of course he's reading a book—I can make out the title, *The Glass-Bead Game*—and of course he doesn't look up from it, not even when Taxi tries a chipper "Hello!"

The background music for this tableau: at least three electric guitars wailing like wounded porpoises, a set of drums being smacked with depressing regularity, and a singer who was born with a nose where his mouth ought to be.

The wavery nasal singing seems to be entering Taxi's bloodstream like sugar. While I watch he hauls a couple of treasures out of the bins with irrepressible surprise and joy.

"Hey!" he says, and beckons me. I sidle over slowly, letting him fidget.

"Look!" he says. "Look at this! The first Moby Grape album—in mono—with the finger cover!" I look. A scruffy bunch of guys who couldn't decide if they wanted to dress up like cow herders or hobos are staring oafishly out at me. One of them has the middle finger of his right hand splayed out on the washboard he is holding. A washboard? And the next guy is holding a harpoon.

"Very mature," I say. "Buster Peeks did the same thing in all the club pictures of the yearbook last year, and they had to reprint half the edition. Nice."

Taxi is deaf to criticism of Moby Grape. "That's what they had to do with this too!" he says gleefully.

46

Then he lowers his voice: "Probably only a couple thousand made it into circulation. *The* seminal Frisco Sound record. *Everything* is in here." He shakes his head. "I had some great music."

"What happened to your records?"

He looks down and resumes flipping. "Your mother kept them."

This kills the conversation for him, and I use the chance to get away to the sofa. I sit and stare at Ponytail. The record finishes playing. Ponytail stands up, reads a last sentence, puts his book down, and takes the record off the turntable. He starts to flip through a small stack of albums, then looks over at Taxi.

"Hey, man," he says. "Is there, like, anything you would dig to hear, man?"

I laugh. He's kidding, I'm certain. But my laugh falls flat, and Taxi gives the guy a wimpy smile, pauses, and says, quite seriously, "I'd really like to hear the first Procol Harum album."

"Far out," the guy says, somberly nodding deep approval and plucking a record out of the pile. "Got it right here, man. Listened to it an hour ago—must have known you were, like, coming. Wouldn't think it would be good for the morning, with all the organ and stuff, but this is great freaking morning music, man."

Suddenly the speakers shudder, and the cat springs up onto my lap. A ham-handed pianist stuns a tinny upright with a bad mockery of Rachmaninoff, and a

47

couple of bars later an organ wheezes out a bad mockery of Buxtehude. An electric guitar kicks in, and then
a voice that should never be allowed within ten yards
of a microphone wails:

> *I am a sad, six-triggered bride*
> *Searching for a place to hide . . .*

"Try the Rockville Home for the Deaf," I say, but
of course nobody hears. The organ doubles its volume
for a solo; the cat burrows into my lap. I'm allergic
to cats. I spill this one onto the floor as gently as
possible. Ponytail notices me dumping the animal and
gives a concerned frown. He rummages behind the
counter and comes up with a green gallon jug, half
full. There is no cap on the jug.

He holds it out to me. "Want a hit of wine, man?"

I snort. Nobody has ever called me "man" before.
"Are you talking to me?"

He nods. "Sure, man. Do you good. You feel to
me like you could, you know, use something to like
mellow you out a little."

I shake my head in disbelief. He takes it as a refusal
and gives a compassionate shrug, as if to say, Hey,
like, I tried. He holds the jug out to Taxi. "What
about you, brother? Some grape?"

Taxi smiles sweetly and says, "No, but thanks,
man. Really."

No, but thanks, man, really? Who said that? Not my

father, the prize-winning newsletter editor who has gotten almost as many good reviews for his vigilant prose as I have for my pizzicato. I groan aloud in disgust. The only thing left to do is pick my calluses and hope we go soon.

Taxi darts and dashes for about another twenty minutes, crowing when he plucks out a masterpiece, flashing me a cover now and then and saying things such as, "Buckley was doing things in vocal music that were dead even with guys like Dolphy in jazz" and so on. Soon he has a stack of about twenty-five records, which he lugs over to the couch to show me, one by one.

The covers blur by in Taxi's hands. Happy Jack. Sergeant Pepper's. Buffalo Springfield. *Dusty* Springfield. Double Dynamite. Blonde on Blonde. Grisgris. Chocolate Watch Band. Love. Fugs. Animals. It's like a bad poem. My head spins with colors, faces, logos, hair.

Finally Taxi gets up and jounces over to the counter. The piano-organ-guitar tirade is just ending; Taxi indicates that he'd like to buy that one too, hot off the platter. Wonderful. As a quick antidote, I try to remember the first few bars of Bridge's Cello Concerto. I can't. Shit, have I already been brain damaged by this noise? I've read articles about this.

Ponytail is going through Taxi's stack and looking up each item in a well-thumbed notebook, jotting things down as he goes. He makes little comments on the selections, like "Wow!" and "Hey, great, man!" as if he were a pal flipping through Taxi's records at home.

For a couple of minutes the guy is silent, adding up a row of figures once and then again with lip-moving, brow-wrinkling concentration. He finishes, reads the total to himself, sighs, reluctantly puts down his pencil, and looks up at Taxi.

"Now the stone bummer, man," he says, frowning miserably. "Before I like lay this number on you, let me communicate, man, that I'm like very opposed to dealing with this whole money trip. These things, man, this music, should be free. . . ."

Taxi smiles and shakes his head, holding up his hand, interrupting the guy's protest with "Hey, look, it's okay, we all have to get by," and "It's value for value, man, and that's cool," and other idiotic things. Finally he gives up trying to prepare Taxi, who smiles magnanimously; the guy sighs, wiggles his hairy eyebrows, and says, "Okay, man. Don't say I didn't, like, warn you. The rip for these comes to $365."

Taxi's smile freezes; his face looks like a photograph of his expression a second ago, gone from color to black and white. "What?" he says. "What?"

The guy is slinking under woe. "You get a free bong for going over three bills," he tries.

Taxi cannot speak; his eyes flit from the records to Ponytail's face to the turntable behind the counter to the records again. All he can say, one more time, is "What?"

Ponytail smacks his hand down on the counter with surprising violence, hangs his head and shakes it, and meets my eyes sideways. "I'm going to quit this freaking gig," he says.

"Oh, I really doubt it, Bonzo," says a new voice.

It belongs to a snappy little guy in a dark-blue suit, standing by the stairs with a stack of records under one arm and the other hand cocked on his hip. Ponytail looks up and says, "Man, listen, like I was only . . ."

"No sweat, Bonzo," the snappy man says, looking at us as he walks over and claps Bonzo on the shoulder. "Your childlike indignation is part of what I pay you for. Tool of the trade for the true hippie. Now run and cross-index these goodies I picked up at a garage sale in Potomac yesterday."

Bonzo nods and slumps away through the bead curtain. The new guy continues to stare at Taxi, who just stands there looking glum.

"So, Little Keats," the fellow says, "what have you been up to for the past seventeen years?"

Taxi's head snaps as if he'd been stung on the neck by a wasp. "What did you call me?" he says.

The man laughs. "Little Keats. Little Keats the English major. When he's stoned, likes nothing more than a good piece of mystic Romantic doggerel to read by the light of a patchouli-scented candle."

Now, I know Taxi used to take some stupid drugs when he was in college, because we talked all about that stuff in the seventh grade when P. J. McCullers got caught selling cow tranquilizers to kids at lunch. But Romantic poetry by candlelight—this is news. I grin at Taxi, but he's too surprised to notice. I can see the whirring behind his eyes as he stares, though. He's about to get it. "You just can't be."

The guy gloats. His hands are still on his hips. I never saw anyone keep them there so long; he must have read that it makes you look tall. "Try me," he says.

"Our on-campus, corner-culture drug dealer. Our man with the psychedelic pipeline." Taxi shakes his head. "But I don't remember your name."

"Of course not," the guy snaps. "You never knew it." Then he laughs.

Taxi nods. "Nicknames. Yes, you got us to call you by something different almost every time we saw you." He shakes his head again. "I really can't believe you're . . . here."

"Instead of in jail?" the man says, and laughs. I think I'd rather hear Procol Harum again, at twice the volume, than another laugh from this guy. He goes on: "You thought I'd be locked away in Mexico or somewhere, right? You can admit it. Hey, I started that rumor myself. But while you thought I was eating bananas for dinner in Mazatlán, my friend, I was streaking through Stanford Business School and a few other places that might amaze you."

Taxi doesn't look amazed at all; if anything he looks as if things are starting to make sense. "Of course," he says. "Stanford Business School. Hm." He pauses. "To think you used to come on like a cross between Mick Jagger and Che Guevara. When you sat with us smoking dope we thought was better just because you were so hip. And we ate it up." He sighs. "I suppose I should applaud belatedly for some excellent guerilla theater."

"Theater is not the word, amigo. It was *marketing*. Selling mediocre pot to college boys was a very easy way to make a buck. And competition was high. I found an edge and stayed ahead." He smiles. "I was an artist of the times, Keatsy."

"A con artist."

"Nonsense. The drugs you paid for were what you got. I never gypped you."

"I'm not talking about the drugs," says Taxi.

"What, then?"

Instead of answering, Taxi asks: "What new market did you move into?"

The man shakes his head. "Let's talk about you for a while, Keatsy. Let's see. . . . You were just a freshman, but I bet you stuck with English, right? Got the B.A.?" Taxi nods. "So—what have you parlayed that keen grasp of Samuel Beckett and Henry James into? Besides lots of excellent conversation, I mean."

"I publish a newsletter."

"Ah," the guy says, leaning forward a little more. "Which one?"

Taxi blinks. Usually, people just say, "Oh, how nice," and move on. "Well," he says, "it's called *Environmental Impact.*"

"A weekly publication in which our hero laments the disgraceful pollution of our precious natural resources by wicked, profiteering creeps who carelessly dump baby-crippling toxins into every estuary and eddy in the biosphere." The man grins. "I know it well."

Taxi says, "Nice to meet a fan."

"Liberal horseshit," the guy says benevolently. "But you're good, Little K. I can't believe what you dig up sometimes"—he laughs—"and neither can some of my clients."

"And who are your clients?" Taxi asks mildly.

"Oh, let's just say various traditional American causes. You'd probably call them 'conservative'—at best! Mostly operating through political action committees. I'm what you might call a fund-raising co-ordinator."

"Of course." Taxi looks over at me. "We'd better hit the road."

"Shall I add these up?" the guy says, patting the records.

Taxi shakes his head. "Is this really *your* store?"

Arms-akimbo laughs again. "Yes, it really is. Feels pretty groovy, doesn't it? You can almost hear a good rap about Ken Kesey or Norman O. Brown in the background. Almost smell the fumes of some dy-no-mite *gold*, man, or maybe some Colombian . . ."

"It's a smart reconstruction, yes."

"And isn't Bonzo fabulous? I found him playing a guitar at Dupont Circle, obviously starving, but turning over all the money people threw into his guitar case to the Biafran Children's Relief Fund. He really makes the joint, don't you think?"

"He's certainly its best feature."

"And the records—what a riot! Do you have any idea of the market?" He holds one up from behind the counter. Its cover shows some blond boys with helmets of stiff hair and brightly colored jackets staring at the camera with their mouths closed. "The third record by The Strawberry Alarm Clock. Who would think it? Only six thousand pressed. I'll get a hundred tacos for this baby. Bought it for half a buck in a thrift shop." He grins at Taxi. "Let you have it for seventy-five."

Taxi smiles slightly. "Who *does* buy your records?"

"Your old buddies. The former members of the great brotherhood for peace and love. All those beautiful boys and girls who have grown up and now wear jockey shorts and bras and go to real jobs. All they want now is a toke of nostalgia—a hit of the old hip days, man—when their lives *meant* something. When they had *convictions*." He chuckles. "So I give them the chance to recapture the time when they wasted their life listening to garbage like this."

"What kind of music do *you* like?" I ask. I can't help it; I have to know that about everybody. While he's figuring out what to say, I promise myself that if this sleaze says, *Oh, you know, Kodaly and Bridge and Chausson, and some of the contemporary Americans,* then I will burn my cellos and take up the electric flute.

He gives me a smile he means to be kind of sexy. "Oh, I don't know, any kind my friends and girlfriends want to hear, I guess. Why? What kind do you like?"

"The music of your mouth saying 'Goodbye' and fading into the distance," I say. "That, and The Strawberry Alarm Clock. They're my *most* favorite."

I walk toward the stairs and down them, and Taxi follows. Clump, clump. Before we hit bottom, a crash of music blasts from the stereo behind us upstairs. "Is it . . ."

"Yes," says Taxi. "Your most favorite. Got a hundred bucks? We can go back."

"I'd rather put it in a bong and smoke it." We step out into the daylight, and I head toward the bus. But Taxi stops, listening.

"Do you hear something else?" he asks.

I stop and listen: yes, I hear it too. It's music, very soft music from a plucked set of strings, coming from near the record shop.

"Over there."

We walk down the side yard between the house and the concrete yard of a former gas station that is now paneled with wood and hung with ferns and calls itself Kwik Health Yogurt Bar & Caterer. The music gets louder; it's a guitar. We turn the back corner of the house. "Ah," says Taxi.

There, sitting on a short brick wall along the side of a stairwell leading down into a basement door, is Bonzo. He is playing a beautiful fat acoustic guitar, with his big head leaning close over the body and nodding softly to the lazy beat of his dawdling, bluesy tune. He sees us but doesn't stop, crinkling the corners of his eyes in a kind of smile and slowly blinking them once as a greeting and invitation to listen.

Bonzo can play. His strokes are very soft, and bring from the steel strings a sweetness I've never heard from metal, a firm hold on pitch but a delicate tone. The tune isn't much itself, but his tricks with it are clever and relaxed. I get the feeling he's playing in the endless middle of a tune inside which he spends a lot of his life, letting it go and picking it up when he has the chance, taking it down this path and that, never all the way to the end. It's a lovely tune for that.

While we listen I see Taxi taking a quick concerned look down the stairs past Bonzo's shoulder. I peek too, and see the frayed tail end of a hooked rug in a little bit of light thrown from far inside by either a dim bulb or a candle. A little off to one side I see what I figure out is the heel of a sandal. And then I sniff: incense. Taxi and I look at each other: yes. Bonzo lives here. Poor Bonz.

As if our discovery spoiled the mood, Bonzo begins to fade out of his current run through this song. He finds a nifty little four-note pattern from two of the chords with beats on the one and three, then the two and four, and he notches the volume down smoothly and expertly through a long, patient fadeaway. When he finishes, still hunkered over the guitar, he gets the best response anyone can give a performing musician at the end of a piece: silence.

After a suitable time, Taxi croaks, "Thanks." I clear my throat and thank him too.

"Sure, man," Bonzo says, gently tuning an A. "Glad you could listen."

Not much else to say. Bonzo messes with a couple more strings, but I can hear they don't need tuning and so can he. Taxi is studying the ground.

"Bonzo . . ." he says.

"Michael," Bonzo interrupts.

"Excuse me?"

"My name, man. Michael. Call me Michael. Bonzo is just what he makes me let him call me in the store. It's another little part of his historic preservation gig."

Bonzo/Michael shakes his head and frowns. "I don't know why he decided Bonzo was, like, the ultimate period-piece name. I mean, did *you* ever know anybody named Bonzo back then?"

"No," says Taxi.

"Me neither, man. Never *heard* of the name except that monkey Reagan made a movie with." He gives a bitter chuckle. "Maybe that's the idea. Reagan's monkey. The dude does dig Reagan, like, to a very unhealthy extreme."

"Michael," says Taxi, "where can I get a guitar around here? I'm in kind of a hurry. . . ."

I snap to attention at this one. A guitar? What does Taxi want with a guitar? He doesn't play the guitar. At least I've never seem him play it.

Michael looks at him too, but appraisingly. "You mean, like, to keep? To play?"

Taxi nods.

Michael nods back, and keeps appraising. He says, "Are you any good?"

"No," says Taxi.

Michael nods again. Then he reaches out and grabs Taxi's left hand and holds it in his right hand. Taxi is a little startled, but lets him hang on to it. Michael isn't shaking the hand or anything; instead he seems to be kind of *feeling* it, with his eyes closed, like a fortune-teller in an old movie.

Finally he replaces the hand by Taxi's side, takes

the guitar off his lap, puts it in its case beside him, fastens the case, and without a word holds it up to Taxi.

Taxi laughs nervously. "I can't take that, Michael. That sounded like a *Martin*! Even if I could afford it, which I can't, I couldn't take it from you. You play it the way it deserves to be played, and I'm just a hacker."

"It *is* a Martin," Michael says. "A 1928. I've got a 1927 downstairs. That one's mine. This one's yours. You can play it just fine. I feel it in your *prana* man. You're good."

Taxi shakes his head. "Whose is it?"

"My freaking arm is going to break off, man."

Taxi takes it, but holds it out a little way from his body as if he's only letting Michael rest his muscles for a minute. "Whose is it?" he repeats.

"Well," says Michael, rubbing his right wrist, "technically it belongs to, like, the shop. I took it in trade for some records from a guy."

Taxi is incredulous. "Somebody gave you a 1928 Martin for some rock-and-roll records?"

Michael smiles and nods. "Pretty freaking strange, isn't it? But it wasn't anybody who should have had the ax in the first place, man. It was one of these music hustler types—I don't know if you've ever seen them, but there are *hundreds*. Creeps, man. They don't know anything about music. If it were forty years ago they'd be into like silk stockings and cigarettes, but it's now and they're into rock and roll. They always have something with them, something they can't wait to unload for a picture-sleeve 45 of 'Do the Freddie' or a blue vinyl copy of *Blue Hawaii*. They all claim to just *love* Elvis, man. Next year they'll claim to love Herman's freaking Hermits if somebody'll pay two bills for a copy of 'Henry the Eighth.' "

"And one of those guys brought this guitar into the store?" I ask.

"That's right, man," Michael tells me. "Came in first empty-handed and started digging through the records. That's what they always do, leave their shit in the car, pretend they're just normal cash customers. Pulled out a couple of hot items, some heavyweight trash, man, like a 1910 Fruitgum Company promo, or an original pressing of *Crimson and Clover*—I can't keep up with what's hot at the moment with these guys. Then they always tell you to wait a minute and run back out to the car and come in with something that's usually a lot better than what they're so hot to bargain for. Sometimes it's records, great stuff they can't get rid of fast enough, sometimes it's posters or old Beatles magazines or a gold shoe Little Richard threw into an audience somewhere, sometime. Once in a while it's a musical instrument." He shrugs. "This guy offered to swap me this Martin for maybe four records. Begged me, man. What could I do?" He shrugs, palms up. "I owed it to the *guitar*, man."

"Does that asshole up there know about it?" I ask.

Michael shakes his head. "Couldn't tell him. He noticed the records were gone when we did inventory. I told him they must have been ripped off. But don't worry about that—he claims them from his insurance company for like $150 each. Sometimes he claims that kind of money just because a record isn't selling at his price, and then he unloads it for a little less anyway. Don't worry about taking anything away from that dude, man. He makes sure he's all right."

Taxi looks at me inquiringly. I nod. What the heck, man.

Taxi swings the guitar case to rest at his side, and sticks out his right hand to shake Michael's. "Thanks," he says.

"Yeah," I say, patting Michael on the shoulder. It's surprisingly bony. "Thanks a lot, man."

"Peace," says Michael. "Remember: peace and love."

8

It's about twenty minutes before Taxi and I talk; we've already hit the Beltway and swung around past Mormons-to-the-Moon, which is what I call the tabernacle that looks like a spaceship outside Wheaton, and headed northwest on 270. Washington is behind us; we're on the road.

I'm in the front seat. I feel a pull to get on my stool in the back. Already that's my spot in this bus. But before I go I want to find out a couple of things from close up.

"So," I say, "what's the name for the Age I'm studying? Something about Aquarius?"

Taxi smiles and blushes. "Oh," he says, "Yes, I did say that, didn't I?" He nods and sighs. "The Age of Aquarius. That's a nickname for the . . . the, I suppose, era that started in the late 1960s. In a certain part of the . . . culture. Young people. People just before, in, and just out of college, mostly. Although

it was supposed to be an era that included everyone—
that brought about a big change in the way the world
worked. A swing toward certain values, fresh but
universal values . . ."

"Peace and love."

"Exactly," he says, nodding crisply as if I've come
up with a very brilliant analysis. "Peace, love, gen-
erosity, truth . . . a lot of words that had become
clichés . . ."

"I've got news for you, Taxi."

"Yes, I know, they're still clichés that don't seem
to mean much. But for a while they *did*. Really. You
could say those words and feel precise and powerful.
As if you had discovered them. Or were celebrating
them."

"Were they as important as money is now?"

He sighs. "God, it's sad, isn't it?"

"So the Age of Aquarius started in the late 1960s,"
I say, in my best student voice. "When did it end?"

Taxi looks back at the road. He thinks for a while.
"I'm not sure, really. It all kind of seemed like one
big morning, and in the morning you never think . . . I
mean . . . well . . ." He laughs, puzzled. "I suppose
it *has* ended. . . ."

"Maybe when Bonzo dies it will end," I say. "Maybe
he's the last survivor, like the last slave who died a
couple of years ago. Maybe until then it's just very
late—almost midnight."

Taxi snaps a Deep Probing Look my way. "Maybe that's it," he says.

"Until Bonzo dies, or goes to business school."

Taxi smiles: "I don't think Bonzo will go to business school. Do you?"

"Frankly, I don't give a C-sharp-minor one way or the other, Tax. It was just a joke, you know? Not a serious subject for debate." He looks back at the road, stung. I go on. "So why am I suddenly enrolled in *History of the Age of Aquarius—The Morning Years*?"

Taxi takes a big breath. Here it comes. "Well, it has to do with your mother."

"Oh?"

"Yes. See—there are some things I think you ought to understand before you meet her."

"Because of what she's like now?"

"Well . . . not so much that. I mean, I haven't been in touch with her for years, you know. I only know where she lives because I mail her the newsletter and get address corrections from the P.O. I can't be sure what she's like now, though I can imagine. This has more to do with what she was like sixteen years ago."

"So how are the Buffalo Springfield and Chocolate Watch Band going to help me understand?"

"They're part of a time. An atmosphere. A world. A world in which your mother lived. In which her decisions were made. Think of it this way: If you took a little bit out of an orchestral passage and played it

65

solo, a fragment without the surrounding melodies and counterpoints, it might sound . . . it might be unfair to look at it that way, right?"

"It depends on why you were looking at it. Are we talking about judging its absolute worth or something?"

He laughs. "With you, Sib, we are always talking about judging something's absolute worth."

He's got me on that one. "What about you, though? You say, 'A world in which your mother lived' as if you weren't there."

He nods as if it's a good question, but one he's thought about already. In fact, I suddenly see that he's thought about all these things, no doubt in the last few days. Or maybe for the last sixteen years.

"I lived in that world, yes. And the things we'll talk about, the decisions and so on, they were mine too. Definitely. But I wasn't as . . . avid an exponent of as many things as your mother was, and I'm not as representative of those times. Being around me wouldn't necessarily teach you what you need to know."

"The way being around Bonzo would."

He considers this. "Maybe."

"Okay. Say you and Bonzo started out back in the morning of the Age of Aquarius pretty much the same. If I compared you now, and you're where *you* are and Bonzo is still saying 'man' and has a ponytail and says 'Peace' instead of 'Goodbye'—then I could see how you've changed, and what's missing to make you less of a prime example? Is that it?"

"Yes. Brilliant."

"Is my mother like Bonzo?"

He smiles. "That's not it, no."

We think for a while. Then I ask: "Was that crummy music really so important?"

"It was essential!" he says and sighs. "Maybe I should have spent the $365."

"Never." I move into the back, and very carefully start to get my cello out. "Maybe I don't need the actual music *of* the Age of A, though. Maybe we can do something else."

"What do you mean?"

I'm sitting with the cello in position, and the stool feels great. I bow out a long C. "I mean, let's write some music of our own."

"You mean some rock songs?" Taxi says incredulously.

"No. Yuck. I mean"—and here I bow out a couple of choppy notes—"something more like a tone poem. A tone poem to the Age of Aquarius. The Morning Years." I play the "Morning" theme from *Peer Gynt*, heating it up a little, ya dadadada dee, ya dadada dee dee dee, ya dada ya dada. . . . "What do you say?"

"I don't know. . . . I'm no musician, as you know."

"You don't have to be. Look. It's easy. You throw ideas at me—causes, foods, fears, anything—and I turn them into sounds. Then we arrange them into a composition. We're working pretty roughly here, but we can swing it." He says nothing, so I play my ace: "Do you know who was probably the greatest American composer of tone poems?"

"Who?"

"Duke Ellington. Your hero. So come on, try it." He smiles. "All right."

I tighten my bow and whip through a couple of weird scales. "Here's how we'll do it, so it's fun for you too. You think orchestrally—that is, you choose what instruments would play each part. And how they sound. That makes it better for both of us, okay?"

He nods. "Sounds good. Let's start."

"Fire away."

He sits and thinks. "Something first from the violins," he begins. "A small statement. Something tuneful, bright but not at all profound, maybe almost even cute. Not too interesting, either. Something . . . something that has been going on for a while, but is getting closer to the surface."

"Phew! You like really know how to challenge those poor freaking violins, man."

"Okay, okay—try this." He's getting excited, tapping his index finger to his lip while he thinks. "Just give me the equivalent of antiseptic kitchens in which crew-cut husbands eat lime Jell-O with banana slices and miniature marshmallows suspended in it, served by perky housewives who got the recipe from a television commercial earlier in the day."

"I get it. The Fifties. Pretty clichéd description though, isn't it?"

"The Fifties *were* a cliché. Play it."

"Yes *sir*."

I think for a second with my left fingers hopping on the strings; here goes: a little something high and sweet, a little jingle, maybe part of a television-show theme.

Taxi smiles back over his shoulder. "Perfect." I play it a few times, adding an accent where needed, until I have it bouncing. "Amazing. This is fun," says Tax.

"What's next?"

"Okay." Taps his lip. "We get the little tune for a while, just long enough to lull us a bit, like the beginning of one of what you once called Mozart's 'throwaway symphonies.'"

"*Movements*. I said 'throwaway movements.' Mozart couldn't have thrown away a whole symphony if he tried. Which he did try. That's genius—you can't screw up even when you want to."

"Anyway—we're getting lulled."

"Lulled." I lull, bowing with a heavy hand, slowing down the vibrato to an even pulse and ebb.

"Great. We're sitting back. A pleasant, shallow melody. And then."

"Pow?"

"No, not pow. Something from the oboe. A little nagging at the borders, a little hint of a tune that could grow to haunt us. The violas and cellos cover it up

69

quickly by harmonizing immediately with the original melody."

I continue the first tune in the middle range with the big vibrato, but then all of a sudden jump to a high pitch for a sharp, shrill line. Taxi shakes his head.

"Too birdy. Too European. Make it more . . . Eastern. Mysterious. Hidden but stinging."

"Maybe if you just told me what it *is* . . ."

"Vietnam."

"The country?"

"No. Yes! Of course. And a lot of other things. The war—a lot of hot, green deaths over there, and over here the rallying point for all of the divisions—right–left, old–young, straight–hip . . ."

"How's this?"

From the first melody I finesse a childish motif up high, five notes I separate into waltz time, the last two notes reversed in every third reprise. I know I've got it. Taxi isn't breathing. "All *right!*" I whisper, and he nods once.

I play it for a while, softly, and then fade it into almost silence; then *hwawww* I bow back the first candy theme, making it sound pompous. From time to time at a rest I pop a couple of those oboe notes back in.

"Next?"

Taxi swallows. "Wow. Okay. Now that nice little melody starts to look . . . well, not just wrong, or silly, or wicked or naive or pointless—but as if it

simply isn't enough anymore. It needs to change."
He holds up a hand while he thinks, so I won't start
until he's finished. "There's some harshness and satire
involved here, but mostly it's the growth of a new
melody."

I groan. "And I thought I'd have trouble drawing
anything out of you."

I play the melody through a few repetitions, using
the Vietnam digression, and then begin to bow closer
to the bridge, thinning the tone, but at the same time
I exaggerate the melodrama of the tune, with silly
sustains and almost comic vibrato. A nice tension:
something's being lost.

"There goes the last of the Fifties values," says Taxi.

I keep thinning and straining, but then very gradually I shift a pitch here and there by a quarter tone,
little by little changing the notes. Pretty soon the pitch
shifts have taken the melody over from the inside,
and it's a new tune altogether. This should be a change
you don't notice—suddenly the old melody's gone,
that's all.

"Jeez," says Taxi, "*that* came out of nowhere." I
guess I did it well. "That's perfect. See, the Sixties
came out of nowhere too. One day you were sham-
pooing twice a day to make your hair look like vinyl
and wearing starched shirts and alpaca sweaters and
doing the Shingaling in front of a full-length mirror
and going to mixers . . . and the next day you were
wearing sideburns and loose jeans and a quilted long-

underwear shirt and listening to music you could *never* have done the Shingaling to. Yep. It was sudden. You're getting it right."

"Is there more?"

"Oh yes." He laughs. "I'll throw things at you pretty fast now."

"Let's go."

"Item: a new exploration of—and trust in—sensuality. Not just sex: anything that feels good should at least be tried."

I insert a new figure in the mid-bass, vocal and vibrating.

"Item: The feeling that there's more to perceive. Not looking for answers, looking for *looking*. Add in the reckless conviction that anything that brings a new perception won't hurt you."

"You mean drugs," I grunt. The melody and mid-bass figure keep going, but I slide into some high glissandi on a couple of selected treble notes in the theme; the effect is pretty eerie, like someone almost going nuts but not quite.

"Ugh," shivers Taxi. "Psychedelic to the life."

"Anything else?"

"Item: Getting high and laughing too much and too loud and dancing in public places, usually in the sunlight."

Easy. I lift a part of the melody up an octave into a 4/4 happytime folk-dance twirl, and integrate it with the other figures.

"Good. My feet twitch."

"Next? We're getting pretty patchwork here, Tax. Is there much more?"

"The Age of Aquarius was pretty patchwork," he says. "Item: a distrust of everything big and powerful."

I break one of the longer melody notes into thirds and play it that way every time the melody repeats. A smallness, a nervous smallness, in the middle of things.

"Item: A growing taste for political aggression, including a questing after the power we distrust."

I keep playing all the melody and figures, but then in the middle I drop them completely for a few seconds to chatter out a passage alternating a single-string voice with double-stop pops. Then I pick the melody up again on the beat. Agitating.

I play the whole thing through a couple of times, starting at the point of the complete melody change; then, because Taxi isn't adding anything new, I go back to the very beginning and move through the entire piece, thinking it's pretty snatchy but still interesting—pretty weird at times, but overall kind of buoyant and fun. I take it a little slower and play it through again. The slower tempo is better—it lets the piece say "man" and "like."

Taxi is nodding up in his seat, beating one hand to the time of the main melody, even humming off key as he recognizes this and that figure. We go through it a couple more times, and I think it's time to wrap it up. "Okay," I say, "let's finish."

"I think we've done it," he says.

"Oh no. This isn't finished. This is a nice, big, thematic salad, but we need to put it in a bowl and serve it. We're scattered all over the place. We need the big perspective, Tax. The unifying factor, the polish. You know."

"No," he says. "I don't. Can . . . can *you* think of something? Something technical that pulls it all together? Something musical, I mean?"

"Oh no, you don't. That's not how we've been working. That's your job. Come on! Just give it to me! Where did all of this wonderful stuff lead? What happened to it? In real life, Tax—just tell me and I'll work it in."

He's absolutely blank. "I don't know, Sib."

I stop playing with a *twong* on the D string and gawk at him. "But you've been here the whole time— I mean, you've lived through it, through whatever happened. From Bonzo to Reagan. I mean, just *tell* me. What happened?"

He stares out at the road. "I don't know. Yes, I've been here all along. But pinning it down . . ."

"But this stuff isn't still going on. What happened to it?"

"Maybe it will happen again." He sounds like a kid on Sunday night hoping his school will disappear by tomorrow morning.

I pack the Bianchi away, snapping the clasps loudly. "You don't believe that, Tax. And I don't either. And all we've got is a messy piece of half music."

He sighs. "I'm sorry. I just want you to get an idea, not a judgment, but an understanding . . ."

"Oh, I'm understanding. I may not be learning what you anticipated, but that's the risk of open education, right?"

He says nothing. Neither do I. I sit on the stool in the back, and watch western Maryland go by through the curved window of 1965.

9

The only thing I know about my mother is from this stupid game I used to play. When I was eight I got crazy about lists. If the kids in your reading class were animals, what would each of them be? What car would each of them be? What tree? I made these lists up constantly for a couple of weeks, and exhausted all the people I knew at school. Pretty soon the only subjects I had left were Taxi and myself, and when I was making a list on the two of us, I suddenly had the weird idea to include my mother. I figured she would fall somewhere in between Tax and me, and guessed. I didn't make up a mom of my dreams; I really tried to fancy what she could be like.

Here's my list:

CATEGORY	SIB	TAXI	MOTHER
Animals	Spider monkey	Otter	Cat
Buildings	Weird new house	Old diner	Monument
Musical instruments	Drums	Clarinet	Violin
Centuries	20th	19th	Middle Ages
Games	Poker	Scrabble	Clue
Colors	Dark brown	Green	Purple
Times of day	Morning	Morning	Night
Seasons	Spring	Fall	Summer
Trees	Maple	Spruce	Birch
Flowers	Iris	Tulip	Rose
Languages	French	English	Eskimo
Cars	Volvo	Volvo	Cadillac
Countries	France	England	Iceland

I made out a list with the categories filled in, the names named, but the lineups left blank. Then I gave it to Taxi, explained the game, and asked him to fill it in for me, without telling him my mother was on it. I wanted to see if he'd faint or something. He took the list, filled it out very carefully, and gave it back to me the next morning.

CATEGORY	SIB	TAXI	MOTHER
Animals	Racoon	Seal	Cat
Buildings	The Villa Savoie	Used-book store	Health-food bar
Musical instruments	Cello	Oboe	Harpsichord
Centuries	20th (late)	20th (early)	18th
Games	Baseball	Kickball	Acey-deucey
Colors	Chestnut	Khaki	Bone
Times of day	Late afternoon	Morning	Midnight
Seasons	Spring	Fall	Spring
Trees	Hemlock	Beech	Palm
Flowers	Columbine	Daffodil	Wild rose
Languages	French	English	Chippewa
Cars	Alfa Romeo	VW bus	Restored Packard
Countries	U.S.A.	England	Middle Earth

Can you figure out what the woman is like from this? But that wasn't really my point. I showed Taxi my list. He read it, handed it back, and said: "You were right about Clue. I was right about the cello.

And I've always wanted to be an otter, but didn't dare flatter myself. Thanks. This was fun."

This was fun. Sometimes the guy is a younger kid than me. Sometimes that isn't so bad, though.

For one thing, Taxi never "brought me up" like any other adult would. I have grown up myself. It sounds cocky, it sounds ungrateful, but it's true. Taxi stayed out of my way. He never protected me from things that were coming up, he never tucked me under his arm and told me everything was all right. He let me find things out for myself. He got me clothes and food and lined up schools and all that stuff. But I directed myself. Don't think bad of Taxi for this; it's the best way. Look, I've seen some classic fathers, and I'll take Tax any day. I'll give you an example. I once was starting to make friends with a girl named Audie a few years ago. She was smart, and had a lot of energy, and wasn't silly. I thought she was probably going to turn out to be okay.

She asked me one Friday if I wanted to spend the night at her house. It meant skipping my evening practice, but once in a while I do that if something comes up. That night the two of us and Audie's father played a game called Parcheesi. Moronic. You roll a couple of dice, you move your little plastic teardrop in one direction the number of spaces indicated on the dice, and if you get to a certain place ahead of everybody else you win. No real strategy, no clever choices to make. Who cares? I didn't; I was bored to the point of memorizing the lyrics of the musicals Audie's father kept putting on the hi-fi, which the two of them would suddenly pipe up and sing along with every once in a while.

Anyway, it took me over an hour to notice that Audie had won every single game of Parcheesi. Four or five in a row. If it had been a game of skill I could see being unbeatable. But it was a game of chance— and chance doesn't usually line up that neatly on one person's side. This didn't seem to bother Audie; she crowed and clapped, and I had to chalk her up as a twit by the second victory. I needed something else to get interested in, so I decided to investigate.

It took about two minutes to discover that Audie's father was cheating. He always cut his own moves back a space or two—if he rolled a nine he moved seven. And he nudged Audie ahead now and then by the same little percentage—an eight got her ten steps forward. Sometimes he gave himself an extra space, if the lower number he rolled would have put him in a good spot, and he did the same when her full roll would have put her in trouble.

I was astounded. Whenever Taxi and I played games, he beat the crap out of me if he could. I didn't get all crushed and sulky; I got curious about how you went about winning. And I started to watch his moves and figure things out. He didn't treat winning as any big deal, either. He didn't cock-a-doodle-doo about it. If I asked him what he had done to beat me, he explained: "You never put down your face cards, so I always go out early and catch you with them" in canasta, or "You never worry about leaving double-word spaces open at the tail end of words you spell, so I save my big letters until you make a spot for me" in Scrabble. Eventually I figured out my own strategies and pulled pretty even with him. From that point on we beat the crap out of each *other*.

Naturally I couldn't sit back and let Audie and her daddykins sucker me. Besides, to tell the soft truth, I felt sorry for Audie. I guess I thought I was lucky having a father who didn't mess around with my chances, and I thought she was stuck with a dud. So for the last game, I corrected the cheats. When Pops moved five for a six I'd say, "Oh, wait, you have one more left," and push his piece one more with a big smile, very helpful. Once or twice he made a point of starting a funny story when he picked up the dice, . arriving at the punch line when he rolled and moved; twice he told the story and when it was over passed the dice to Audie without moving, pretending he'd taken his turn already. I never let this go, of course.

What happened? Audie got very crabby. The father got desperate. The game crept along in misery. Finally—despite everything he could do—the father won. I clapped and slapped him on the back, congratulating him mightily for breaking Audie's streak. I was a little cruel, but what the hell, he deserved it.

I should say *they* deserved it. They deserved it because the two of them were in it together. Audie knew very well he was cheating for her, and yet she got excited about her victories anyway, the little dip. And he felt like a big benevolent dad, taking care of his girl.

The Audies of the world can keep their daddies' benevolence—I'll take Taxi, even up, no favors.

10

We're driving through Ohio. It's getting dark. We just ate some sandwiches Taxi packed—mine was ham, Taxi's was tofu fried in soy sauce, if you can believe it; he's a vegetarian. Taxi says we'll eat a late dinner when we stop for the night, if that's okay. Sure.

Ohio is dull as a record of madrigals. Almost without thinking, mainly to fill up the time, I ask Taxi:

"So when was the last time I saw my mother, and why was it the last time? Just give it to me straight, Tax," I say, to keep him from trying to get tactful, which he does about as well as he gets sarcastic. "No need to protect the innocent."

But he surprises me by jumping right in, easy as can be. Then I realize: of course he's ready—this is one of the big questions I'm supposed to have been gnawing on for years.

"You last saw your mom on January 1, 1970, in San Francisco. You were about twenty hours old."

"Fill me in. Set the scene. Pretend I wasn't there."

He laughs, then gets serious again. "Your mother and her midwife and you and I were all sitting outside behind the house. It was a pretty warm day for that time of year. Your mother was wearing a big wool caftan and no shoes. It was a purple caftan. She liked purple, but she hadn't worn it while she was pregnant."

"Why?"

"Because one of the Indian tribes she studied said it wasn't proper. I don't exactly remember why, or which tribe." I try to form a question about this Indian stuff, but he goes on before I can come up with anything.

"Your mother was in a chaise longue. Looking pretty beat, but queenly. I sat on the ground. The midwife stood behind your mom, nervously flitting around. She did that during the whole delivery, too." He pauses.

"You're forgetting somebody."

"No, I'm not. I was trying to recall the term for what you were done up in. It's not 'papoose,' though that's the word that springs to mind. In any case— no pun intended—you were bound up inside a leather sack with some beads sewn on in a significant pattern denoting something I don't think I even knew then but cannot recall now, anyway—sorry—and hanging on a stick."

"What do you mean, on a stick? She stuck her twenty-hour-old baby in a bag on a stick?"

"Well . . . It was, as I said, some special Hopi or Cree or something way of giving you a place in the midst of the air and earth elements. Letting the baby rest in the great long pull of gravity, which is a sort of dominant deity or spirit in this particular tribe's system of belief."

"Tax—is there something you're not telling me?"

He smiles grimly for just a half second and says, "Quite a few things, actually—but what do you mean?"

"Is my mother an Indian? Am I a half-breed?"

He looks as if he doesn't know whether to laugh or feel sad; finally he goes for the sadness. "No, you're not a half-breed. And no, your mother is not an Indian."

"Then why do I notice this subtle strain of Indianism running through the discussion thus far?"

He sighs. "Your mother was . . . very wrapped up in a lot of lore from various Indian tribes. She thought Indians were—or, rather, *had* been, before their culture was crushed and reduced to its present insignificance—the most enlightened people ever to walk the earth."

"How did she get into Indians? I mean, it's not the most normal fascination in the world, is it?"

"Well . . . actually, it *was* a pretty normal fascination back then. The hippies pretty much thought they had found their forefathers in the American Indians."

"Did *you* think so?"

"No. I didn't become an Indian-head. I was really in a minority, being a pedantic hairsplitting sort who disliked the hippies' way of grouping all Indians together into one big antecedent. But your mother was a more normal hippie."

"What exactly *is* a 'hippie,' Tax? Like a 'hipster'?"

He smiles. "Less street smart, less underground. Hippies were proudly naive; they wanted to operate innocently, trusting more to intuition than knowledge and rational planning. They wanted to live in harmony with the natural forces that governed the growth of trees and eagles, rather than the conventions of social structures, which they saw as crass and rigid. A hipster was a city creature; a hippie was strictly for the forests and mountains and rocky coast."

"Okay. So my mother's a hippie, and you're sitting on the good tree-growing earth, and I'm hanging in a bag getting to know the great god Gravity." I shake my head incredulously. "Jeez, Tax. I'm already pretty cranky, and all we've talked about so far is what people were wearing!"

He stares ahead grimly. I have to ask: "So what happened next?"

He thinks. "Well . . . first we had a discussion about the bag. I wanted to hold you, or have your mother hold you. You looked so isolated hanging there. . . ." He shakes his head. "Anyway, your mother said that you wouldn't be bonding to us for three more days, so you might as well bond to nature. For that period of time you couldn't even see us, and we would just be distractions from what you *were* seeing."

"Which was?"

"Angels."

"*Angels?* You mean fat babies with wings, playing Vivaldi on harps and lutes?"

"I gathered from your mother that they were more along the lines of flashes of pure light."

I am gaping. "And this was all I could see, supposedly?"

He nods. "For about three days."

"Was this cosmic optometry a piece of Apache wisdom? Or was it Cherokee?"

He wrinkles his forehead. "I think, actually, that the angel thing was Vedic in origin. Your mother was interested in a lot of translations from the Vedas of India. It wasn't just American Indians. She . . . picked up a lot of things from a lot of sources."

"I bet. So—during this time of angels flocking around my head, I wasn't to be bothered by my mother and father doing things like hugging me?"

He gives me a little hike of the eyebrows. "I'm afraid that's just about right," he says.

"How convenient."

He turns and looks quizzical. "What do you mean?"

"How convenient that no, what do you call it, *bonding* could take place for a couple of days. Makes it very easy to dump your kid, doesn't it? Not like you're really even in the kid's life yet, right?"

"I know it sounds bad, Sib . . ."

"Oh, no, Tax; it doesn't sound bad. It sounds *ludicrous*. Ludicrous is better, because of the entertainment value. Tell me more."

"First you have to understand, though, that these beliefs of your mother's were sincerely held, really, and not because they made other decisions seem easy. There wasn't any cunning involved."

"Oh no, of course not. We're talking *innocence* here. I'm hip."

He presses on. "Well . . . so we had a talk about the bag, and that led to a larger talk about . . . oh, I can't remember, some theory of the soul or something." He takes a deepish breath and slows down. "Then I couldn't stand it anymore, so I went over and took you down out of your bag. It took me a couple of minutes to unlace the thing, and during that time I heard your mother talking in a low voice with the midwife. When I turned back, the midwife was gone, and your mother was meditating."

"Meditating—you mean TM, like you do?"

"No, this was a different kind of meditation. Your mom . . . we started TM together, but it wasn't flashy enough for her. She didn't like just sitting down twice a day in private and getting a little deep rest. She wanted something more dramatic. And something she could do whenever she liked. No routine." He hesitates, then says: "She kind of liked an audience, too."

"An audience? For meditation?"

"She would drop out of a discussion for a quick minute or two quite frequently. Close her eyes and cross her legs and make circles with her thumbs and forefingers." He shrugs. "One simply waited for her to return."

"Fun gal at parties, I bet."

"Anyway, she was meditating when I turned around with you. After a couple of minutes she opened her eyes and said she had come to a decision."

I find I'm actually a little nervous here, even though I know what's coming. I want to hear the terms.

"She claimed the decision had been made at dawn, not during her little meditation. A Buddhist practice, I think, making decisions only at dawn. Mind's clearer then." He hesitates. "I said, 'What decision?' And she closed her eyes again. Not to meditate, though—just to center herself in the moment."

I don't ask what this means; I don't want to interrupt now.

"She said, 'I've decided to stay here.' Well, we *lived* there. I said something along those lines, but she shook her head as if I weren't trying. 'Your future lies eastward,' she said."

"Sounds like the Ancient Master in every single kung-fu movie I saw at a martial arts film festival in Munich."

He nods, though he's far back there now, right out on that San Francisco lawn. "It was a delicate way of

89

telling me that she wanted to remain alone in California, and . . . let me take you with me."

"Oh, very delicate," I snort.

"We talked about it pretty frankly, then. But there simply wasn't a whole lot to say—she felt that if you stayed with her, if *we* stayed with her, no one would be happy."

"Did she use the word 'happy,' really? Somehow I can't see it."

He stares at me sideways; I look straight out the windshield. "No," he says, kind of spooked. "No, she didn't. She said 'self-realized,' actually."

"That sounds more like it, man. Go on."

"That's it," he says. "We decided that the best thing for everyone was for you and me to go away together."

"That's not 'it,' Taxi, not by a big Comanche-Buddhist long shot. Why?" I turn and glare at him. "Why? What did she . . . *why*? I want to hear what she came up with."

"Well . . ." he begins, "Well . . ."

"Don't get crafty, Taxi. Just hit the big points and play it straight."

He glances at me, peeved. "A little less of the queen-to-lackey act, all right?"

"Sorry. But this is the story of *me*. You should tell it the way I want it. You're already tinkering with the *trip* for some purpose I'm polite enough to ignore.

This part you should give me without any fooling around—just like it *was*."

"Okay," he says, still peeved. "Your mother wanted to pursue several things in her own life, things that raising a child would prevent her from doing. What you have to understand is that back then . . ."

"What things?"

"Back then, it was actually *more* moral to base your decisions on . . ."

"*What things*, Taxi? Cut the dodge. What things did she want to pursue?"

This is it; this is what he's been trying to soften me up for. I can see it in the way his hands stop fidgeting and sit on the steering wheel like a pair of old doves finally resting, not bothering to retreat when someone sits on a bench too close.

"There were several things," he says. "First, her . . . spiritual studies. Her quest for self-realization. She had gone to a few seminars and things, spent a week here and there at institutes like the one you were named for. She thought she needed to be free to go to more, to follow wherever her studies led her."

"What did her 'studies' involve, Taxi? Study, the way I study cello?"

"No," he says. He shrugs. "She read a few books."

"Second question: why didn't she 'realize' before I was born that she was going to need to do this? Why was it a surprise all of a sudden?"

91

He's really frowning now, trying to work out how this will sound. "Well . . . there was a system of belief—Indian or Vedic or Buddhist or something—that said a woman attained her highest state of being when she gave birth to a child. Your mother took this to mean, I don't know, some very clear change of consciousness. And, you see, it hadn't happened."

"You mean I had already been here for nearly a whole day and she didn't have X-ray vision yet? Her shit still smelled and she couldn't translate the song of the chickadee into English and was completely unable to visit Jupiter by ESP? Poor thing."

He smiles in spite of himself. "Then there was her . . . career." He blushes.

"Oh, come on, Taxi. Don't do it. 'Career'? She never used that word."

"Yes, she did," he says.

"What was her career in? Computer engineering? Tax-shelter investment brokering? Geological chemistry?"

He sighs. "Macramé."

"Macramé?"

He nods. "It's a craft, a kind of weaving. It involves a lot of knotting. With rope."

"Rope."

"Yes. She believed that if she practiced a great deal and established the connection between her spirit and her hands, then she could be a . . . well, a great macramé artist."

I laugh. Cruelly. On purpose. Taxi takes it for a second, then says defensively, "Well, she *was* quite talented, as macramé goes."

This is so funny it stops my laughter cold. "What was she quite talented at *doing*, as macramé goes?"

"Owls," he says.

"Weaving *owls*?"

"Macramé is very three-dimensional. She could do a pretty good owl, really. She even worked little sticks through their claws, and showed them perching."

"Why owls? Why not, oh, butterflies or cats? Was there something special about owls?"

"I believe it's because owls are both wise and cute," he says.

"Was my mother wise and cute?"

He hesitates. "Well, she was smart and pretty," he says cautiously.

"Did everyone want to be wise and cute in the morning of the Age of Aquarius?"

"No."

"Well, hooray. She finally did something original."

Taxi says nothing. We ride for a couple of miles. Then I ask: "Are there any more reasons?"

"None that don't sound petty and selfish unless you understand, Sib, that back then one was *supposed* to think first about oneself. It was seen as falsehearted to deprive yourself of your quest. Your quest was your destiny. By fulfilling yourself you would contribute most to the world."

93

"Far out."

He sighs. "It's too much to expect you to understand, I know."

"Oh, jump that, Tax. I can understand the things you're *saying* perfectly well. I mean, selfishness isn't exactly an obscure philosophical concept requiring deep analysis beyond my ability, you know? What I don't understand is how someone can just drop a kid at twenty hours in the name of destiny—were twenty-hour-old kids supposed to be left to fulfill *their* own destiny too? I don't believe *you* really understood it, either, not then and not now. You went along, maybe, and you're sort of defending it now because you don't want me to hate my mother. But I have a feeling you didn't buy it."

He smiles. "That's obvious. Isn't it?"

"I guess it is, Tax. I guess I've learned *something* from sixteen years."

I fall asleep somewhere in Indiana, and Taxi doesn't wake me up until our camp is set up on the western side of the Mississippi River. I slide into my sleeping bag without ever waking up, but I know from studying the maps before the trip, which I always do, that we've gone more than a third of the way. Two more long days on the road and we'll be there. Two more days and we get to see the self-realized squaw.

11

After I heard him on that old recording, my hunt for Dzyga got cunning. I started a file marked "D." I wrote a hundred letters in four months. I opened up even more to the music journalists so that I in turn could ask them about Dzyga and expect an answer. A few of the older ones remembered something about him, but only one—a guy who had spent seven years in Moscow reporting for a London daily, now a music critic for a paper in Dublin—had much to tell me about disappearing inside the Soviet Union.

"Down a bleeding *hole* if he broke a fingernail and queered a recital in front of a capitalist pig at a state show-off affair," he said. "Can't stand for that species of effing failure, don't care who you bloody are. But if he made trouble and talked out of school or refused to wear scarlet on Saint Lenin's Day or some such, they wouldn't put him down so easy. Something perverse in the Sovies—they like to keep the contentious buggers around in the shadows, sad lessons to all, *There but for the grace of the proletariat go I* sort of thing. Maybe show off the castaways once in a while going into hospital or some such pitiful pose. Now someone like Dzyga—top of the world while still on mother's milk—him they'd hang on to for anything short of banging Mrs. Khrushchev. Give him a tiny apartment with no windows and make him practice his Borodin. If he changes his tune one day, then trot him out and pin back the ears of the music world, including the callow likes of yourself. So beware of fifty-year-olds with Slavic accents, luv, especially when they give the miss to vibrato. I heard him once on his wee violoncello, and he's the prince of darkness all right. Send you scampering off to a career in stenography or lovely safe nursing to keep out of *his* way."

The idea that Dzyga might be in danger or even terrible discomfort made me feel like one of those sweet girls in a western watching her nice man get beat up by the rustlers with stubbly beards. I was outraged and sad—sadder, I realized, than I had ever felt about anyone. I had to find him. I think I may have even started to think I had to *save* him.

It wasn't long before I had exhausted books and word of mouth and correspondence again. So I decided to look for the contemporary Dzyga in the same place I had found the historical Dzyga: on recordings. Somewhere he must be on a disc.

Russia and several of the Soviet-bloc countries exported classical records; naturally they wanted to compete well with Western interpretations, especially of Russian music. Would they be able to resist putting their best cellist forward? Well, not even forward—most of the records from Russia were orchestral, so they could probably slip him into big ensembles and even the most astute listener would have trouble picking him out of the swollen string sections playing Tchaikovsky and Rimsky-Korsakov.

My intuition told me to concentrate on the chamber-music discs coming out of Hungary, Poland, and Czechoslovakia. Perhaps Dzyga was farmed out to the sticks. Or—more likely, I thought—perhaps he snuck off now and then to do a studio session under a pseudonym where he was less well known.

So I started hunting with my ears. I had zero luck with hundreds of string quartets, sextets, nonets, sonatas. Once I thought I had him on an exceptional Supraphon recording of a Martinů quartet. The cello part was not amazing or anything, but I could hear a vibratoless sostenuto in several places. I started to drum my fingers and tap my toes: maybe this was it. However, after a few listen-throughs, I suddenly realized that during these parts *all* the instruments were playing without vibrato. Weird. And then I noticed that these moments coincided with the passage beneath the needle of a whopping visible warp in the disc.

But one day I got lucky.

By now the clerks in the classical sections of the D.C. record stores knew me well as a customer with a strange yen for mediocre string music from behind the Iron Curtain. They used to see me coming and unload all their moldy imports on me for eleven bucks a shot, and they must have wondered why I kept coming back.

One day one of them motioned me into his office, asked me to sit down, then went to a filing cabinet and pulled out a good-looking record cover.

"Major contraband," he said in a whisper, with a wink. He held up the cover, a painting of the Warsaw skyline at dawn. "The mystery disc of the month. Maybe of the decade. It's a promo copy of a new recording. Not for sale. We got it two days ago, but this morning the distributor called to say it had been withdrawn by the exporter; he even asked that we return the promo." He shook his head. "There are probably fewer than a dozen copies in the U.S. right now—fewer by sundown if some stores return theirs."

He held it out to me. "So take it and lock it up at night and if the State Department traces it to you, you say you bought it along with some Nicaraguan coffee and Cuban cigars from a congressman's aide. Enjoy."

I took it. From the moment I touched it I knew I had scored. On the way home I wedged into a dark subway seat with my back to the aisle and read the cover.

It was a record of string quartets by three avant-garde Polish composers. This in itself was quite a departure from the usual Soviet-bloc stance that such junk was decadent and disorderly. But there were more departures: the cover was costly, and it actually had photographs, of the composers in relaxed poses and casual dress; there were smart analytical notes inside, translated into French, German, and English; the label had the Direct Metal Mastering trademark, an expensive process; and the vinyl was good, heavy Teldec from Germany. The communists were suddenly trying to show that they had a twentieth century too.

I ran home from the subway station. Taxi was out. I ran into the conservatorio, flipped on the paraphernalia, groaned and jumped up and down while the preamp warmed, and finally put the record on.

The first piece, by the composer Lutoslawski, was maddening. Most of its twenty-two minutes consisted of phrases no longer than eighth notes in single isolated strokes, between long passages of pizzicato. It was easy to pick out the cello, but impossible to know if it was Dzyga.

Side two started out to be just as annoying, in a different way. The Quartet by Penderecki featured throbbing dissonant notes played in very loud unison by all four instruments. That was it, for nine minutes. The cello never stuck its nose out of the throng.

Ah, but the third piece—by a woman named Bruzdowicz—made my heart flutter in the first six bars. *Dzyga*. I heard him, I felt him, I had him.

The composition launched the instruments into long lines played in rounds, and when the cello took its part, there he was, with all three trademarks: the uncanny pitch, the rhythmic oddity and lack of vibrato, and the emergence of each note from that weird silence. I listened with my teeth clenched for sixteen minutes. He was gorgeous.

I grabbed the cover. It gave two recording dates two days apart, almost eleven months ago. I was ecstatic. I was less than a year behind him.

Now what? That's when the letdown came: now nothing. Whoopee—I knew where Dzyga had been on two particular days last year. What could I do to get closer?

Dzyga was playing under a pseudonym: the Debika Quartet listed one Josef Tuza as their big-box man. He was hiding, and only a fool or an enemy would ask any questions that might bring attention—and heat—to him.

I stewed about it for a couple of days. I thought I could make some discreet inquiries among the record critics I knew, but anything I called attention to would get special attention from them, too: one of the jerky things about being famous is that you can't make truly discreet inquiries.

Then the idea came: why not make my fame work *for* me?

I wrote a letter to the record company, saying that a reviewer familiar with my taste for contemporary music had played me a part of the quartet disc. I liked it tremendously. I admired the composers, and the venture as a whole, and I hoped I would not be too presumptuous in putting forth a humble proposal: Would it be possible for me to participate in another such courageous venture, if further records were planned that demanded another cello? If there was, say, a sextet, or a suite for strings, or something of that nature, I would be thrilled.

I wanted to send something along with the letter: an audition tape. At first I fooled myself, trying to believe I was playing the part of a musician trying to crack into a studio date; the tape was just a theater prop. But I knew better, as soon as I set up the microphone at home and picked up the Bianchi. I was making this tape for Dzyga. Whether he heard it or not.

I played the best forty minutes of cello I have ever put together. I played Haydn, I played Bridge, I played Barber, I played short pieces by contemporary American weirdos, I played Bach. And as a special come-on, I spent a week transcribing and learning the cadenza from Szymanowski's Second Violin Concerto, and played that too. When I put down my bow after the last piece, I was pure light and air and sound— I had no body and no identity, just a sparkling joy.

This guy was really driving me.

I sent the letter and the tape.

But then not long after I sent the stuff, the whole Solidarity thing broke; martial law was declared, the tanks rolled, the Soviets marched, the Poles gathered in the streets and threw things. But the Poles who gathered in the streets all looked like dockworkers. What did a cellist do when the iron heel came down on the land he was working in? Hide away in an attic with a couple of suppressed scores, practicing pizzicato passages with a mute on the strings? Dress like a dockworker and hit the bricks with the rest of the guys? Flee?

Now I doubted my letter had ever reached the recording studio. And even if it had gotten through to the studio before the martial law, the response would be cut off.

I gave up. Almost completely.

Then one day I received a letter from California, from the dean of a newly established music school called the Phrygian Institute. It was a nice letter, no big deal at first, though interesting enough. The dean began by congratulating me on the Janáček Prize, naming three of the things I had played especially well during the week of competition, and offering pretty decent bits of analytical praise. Okay, he had an ear.

Then he briefly described the Phrygian Institute. A selective academy with an "extraordinary" faculty, featuring a one-on-one tutorial system for instrumental mastery and group study with eminent composers. Composers?

Yes. This was where Phrygian was hot stuff. It was the Institute's "boldest premise," said the dean, that all great instrumentalists were also gifted with the inspiration of the composer, a gift that no other academy was dedicated to realizing. Too many virtuosi, unable to isolate and develop this compositional urge, went into conducting as a surrogate. Many of them lost their virtuosity as a result. But under Phrygian's system . . . Okay, nice idea. I liked it. But show me some faculty names.

The last paragraph started out like it would cough up some names, but it didn't—the faculty had been assembled from among the world's virtuosi, blah blah blah. It looked like they had nobody. But just as I was starting to crumple the letter I scanned the last sentence, and stopped.

A new member of our violoncello faculty has requested that I invite you to audition for a scholarship for study at the Institute, beginning in September of this year. The faculty, normally out of session in July, will be assembled at 1 P.M. on July 13 to hear your audition. The above-mentioned faculty member requests that you be prepared to play, in addition to your selected repertoire, your transcription of the Szymanowski Concerto cadenza.

12

I wake up in the middle of the night in the middle of the woods in the middle of the country. There is a pine needle in my mouth.

Misery. Being awake in the middle of the night is bad, but what's worse is the taste of the pine pitch. It reminds me of a bad time.

I have forgotten the woman's name. I have forgotten who recommended her to me. I have even forgotten where her "studio" was in town, and I never forget things like that. As soon as my stint with her was over, she was *erased*. I subtracted those days forever, and the first time they have come back, really, is with this pine pitch in the dark.

She was my second cello teacher. My first had died, and because I felt I had been left alone, I had no idea whether I was good or bad at playing the instrument; I had no friends in music. The only kid I talked to about music lessons—a violinist one grade ahead of me, whom I'd seen carrying a violin case through school for three years—told me: "Sure, you've *got* to feel like you're terrible. Your teacher isn't supposed to make you feel *good*. That's the way music works. That's how you learn to play." The next year that girl stopped carrying a violin case to school, and when I asked her about it, she pointed to my cello case and said, "Oh, you'll outgrow it too."

Taxi used to get me to this teacher's place early, so I had fifteen minutes to kill sitting in an ultramodern waiting room beside a bulbous jade plant. I was waiting for the same thing every day: for a door to open slowly, for a fourteen-year-old boy to slink out looking defeated and plucking at pimples on his cheeks, and for a tall woman with a chin like a fist to beckon me impatiently into the room. There were no magazines, not even any music scores to look at. So to keep my nerves busy I somehow got in the habit of sneaking my skinny hand inside the floppy latched case of my crummy first cello and flaking off a few pieces of bow rosin to chew on.

One day after I had walked through the door, squeezing my elbows to my sides and bumping my knees together, and had taken my seat on the chair my teacher claimed came from the back of a first-tier box "at the old Met," the woman sniffed the air with a worried frown, as if a rat had died inside the piano. Then she glared down at me and demanded that I let her look into my mouth. I wanted to punch her right in the knuckly chin, of course, but I held back. Intimidated and furious, I could only manage to shake my head.

The woman said that she was quite wise to my *filthy* habit. It simply had to *cease*. Not only was the chewing of bow rosin inherently *disgusting*; it was also *insanely* dangerous. Rosin was a deadly poison; enough of it to kill me might already have *accrued* in my bone marrow, but if not, I should certainly count myself lucky to have the *opportunity* to desist herewith.

I went home that night and looked up rosin in our encyclopedias. I found that the Phoenicians had prized it as a tooth cleaner, and the Greeks added it to wine to this day. Two of the world's greatest empires either chewed rosin or drank it: the stuff was plainly not molten death.

Even at eleven I couldn't stand the thought of being around someone I knew I couldn't trust. I called the teacher the next morning at 7:30 and told her I was dropping my lessons.

"But, my dear . . ." she sputtered, "do you realize you have the potential to be a player of genius?"

I nearly laughed. I also nearly cried: I missed my first teacher a lot, and the thought of him right then made all the time I had wasted on this woman seem criminal. "Goodbye," I said, with tears she probably thought were for her.

I lie in my sleeping bag looking up through the tree

branches at the sky. Gustavus wasn't a saint, and I wasn't an apostle. Most of the kids at competitions in Europe have these soulful reverences built up toward their first teachers; some of them call them "my master" or even "*the* master," which makes me want to puke into their f-holes.

But, yes, I had trouble finding other teachers after him, to the point that I now use two teachers to cover the ground I know Gustavus would have handled alone. He was a good cello teacher. People who set themselves up as cello teachers are *supposed* to be good cello teachers. Just like people who set themselves up as cellists are supposed to be good cellists.

I roll over in my sleeping bag and look out into the woods. From the maps I got the idea there were only about forty trees in the whole state of Iowa, but here are some pretty decent hardwoods and soft firs. I know a little bit about wood, also from Gustavus. He wanted me to understand how a cello became a cello, so I read a few books and listened to him talk about the making of instruments, and we even took an old cello apart. It wasn't a terrific cello, but later when I learned what even a crummy cello costs, I was pretty shocked that he went that far for a few lessons. He said he was going to glue it back together, but he never got around to it. He was right, though: learning how the instrument worked sort of set a standard for me. I wanted to work just as beautifully.

The first time I met Gustavus was in a back room up the stairs of an old Victorian boarding house in Takoma Park; the first thing he did was hand me a cello and ask me to play it. I didn't want to touch it—I told him I had never even been in the same room with one before, I had just watched the cellist in the National Symphony at one concert (*The Nutcracker* at the Kennedy Center). I was obviously frightened. He nodded, still holding it upright next to a stool I was supposed to sit on. Nodded, but didn't relent. I was simply to try, he said, in his Swiss accent; of what was I afraid?

I was afraid of making mistakes. I was afraid of playing bad notes. I was afraid of starting off with errors when what I wanted to do was start off—and continue forever—with perfection. This was to be the first thing in my life that I had consciously started all on my own, the first thing I had controlled from its start. I wanted everything to be exact and crisp and fine. I knew I had to begin with very basic things, but I wanted to do those things well. I didn't want to scratch and shriek. It would make me feel I had ruined the beginning.

I explained this, more or less, to Gustavus. He watched me closely, nodding, smiling once, and seeming to agree with me. "Of course," he said, when I finished. "You are a very strong young woman, and this is good. So no doubt that you will be able upon this also very strong cello to make these first notes the ones you want to begin with. Do you see? You need not scratch and shriek. But you must begin on your own, not with what I tell you in a lesson to play. It is for both of us a test, you see? Will you like the cello when you meet as strangers? The first meeting is between you and the instrument, not between you and the teacher. The teacher is this time only for observing."

So I took the cello. It felt surprisingly light and swively—I had the idea it would sit like a baby elephant, heavy and placid, while I sawed at it. Instead it spun and twisted in my hands, and I was alarmed— so I spent a couple of minutes finding out how it moved, and how I could keep it pretty steady.

I had thought the bow would be solid and stern, like a hacksaw blade. But it was made of thin wood and fine hair; it was all curve and spring.

And those strings! They felt at first like raw iron. But when I began to squeeze them with my left hand, not trying to pin them to the neck right off, but feeling them firmly, they seemed to warm up and touch *back* a little bit, and when I started sliding my hand up the neck they responded along their whole length. I had the weird sensation that the strings did not stop at the bridge and the head pegs, but started far away in one direction, and ended—if they ended at all—far away in another, just passing through this cello in their course. It was a funny idea, and I laughed. Gustavus asked what I felt, and I told him. He nodded.

After a while I stopped squeezing with my whole hand and started feeling the strings with my fingers, one finger at a time, one string at a time. It probably took a while, now that I think of it, but I wasn't aware of any clock, any more than I am when I play a piece of music now. Once you enter a piece of music, time has nothing to do with the clocks outside; it's not 8:22, it's the fifth measure of the second movement, allegro. A lot of musicians start to lose track of exterior time as their careers go on, to the point that they forget to notice their birthdays and sometimes don't even remember their age exactly.

I was inside musical time that first day at Gustavus's, though I didn't know it. Everything had stopped but my preparation for playing the first note on that cello. And the preparation didn't count—it was using up zero time. I was going neither forward nor backward. The first note was mine to strike, and I would do it and start the clock ticking when I was ready.

I forgot it was a test. What sound would this fingering and this bowing make? What about this sliding, or this angling? And if I plucked the string instead of bowing it? And so on. I had to know more. Before I realized what I was doing, I had started to play. The thoughts of "What would happen if I did this?" had become the experiments themselves. I honestly don't know how long I played before I noticed one finger was numb, and stopped. Only in the silence did I notice what had preceded it: my music, on the cello.

"The test is passed," said Gustavus in a low, sweet voice that I immediately wanted to imitate on the cello. Or maybe he said "The test is past." Who cares? I was in love. I didn't want to stop. He began to ask me questions and I answered in grunts and blurts; talking seemed so stupid and weak at the moment, next to the big round sweet sound that the cello filled your chest with.

Finally, in a very formal way, he told me he would accept me as his student. This kind of surprised me— I had thought I was the one who did the accepting, like I was the shopper and he was the goods. But I just nodded and mumbled something in the way of thanks. He made a little bow. "And now," he said, "I will explain the requirements."

The first few seemed natural, if weirdly worded: I was to practice enough to keep my technique up with my feelings; I was never to practice so much that I lost my curiosity about what the cello could do.

The last requirement struck me as a little strange. Gustavus said: "To me you will bring one composition per week, in your own hand." I knew this meant in my writing. I thought, though, that by "composition" he meant one of the things you wrote slapdash for school, one side of one sheet of ruled paper on an assigned topic, *Why Butterflies Are So Special* or *Where Vegetables Come From*. I had dozens of them at home—Taxi never let me throw out anything I wrote or drew for school—and even if most of them had a big red "C" written on the top, I could recycle them.

But as he talked some more, I got the idea: he meant he wanted me to write *music*. This was the strangest suggestion I had ever heard. I had seen sheet music, in the desks of band students in school, and it had always looked to me like something between the printed word and drawing, a strange new code. When I decided to take up the cello, of course I knew I would have to learn to read that code, but the thought of *writing* it seemed absurd, like when I used to imagine coming up with the chemical formulas Taxi sometimes included in his newsletter to pin down a toxic substance.

Gustavus wanted me to learn how to read music by trying to write it. And I did. The stuff was very crude at first—little five-note "études" that I picked out on the cello at home. Later I got a little more ambitious and wrote longer études; and still later I wrote what I called "sonata movements," which was a pretentious way of making my things sound classical.

When I brought my handwritten sheets of music to Gustavus every week, he would take them with a formal "Thank you," like I had given him a ten-dollar bill. I never got rid of the feeling that when he turned to put the paper in a drawer he was going to give me a receipt.

He never looked at the pieces in my presence, and for many weeks he never mentioned them either. I thought he wasn't reading them at all. But once, after I had proudly mastered a tricky bit of fingering and submitted a piece I had written around a demand for that trick, he wrote a note and mailed it to my home. In it he explained in stern terms that the composition of music worth playing had in its inspiration nothing to do with the technique of playing an *instrument*. Music, he said, played the *human*, and the human played the flute or the drum or the cello. I didn't mention the letter to him, but I remembered it. I still remember it, right to this moment.

My tricky show-off piece belonged to a long tradition between composers and instrumentalists, and this was a good way of being introduced to the ups and downs of it. Here's how composers throughout history have seen it: when somebody's around who can play things nobody could play before, compose something difficult for him so both of you look ingenious. The tradition has produced some great pieces. But more often it lets in hacked-out scores full of opportunity for a soloist to perform the musical equivalent of juggling soap bubbles while riding a unicycle backwards, blindfolded. A few guys wrote junk like this for me after I won at Brussels and Prague. I never played it. But a few smart and talented people have written me music that really takes me places, music Gustavus would have loved to hear me play.

Gustavus's big thing was letting your feeling about a note get into the instrument, and then out as sound. All the technical effects some cellists learn and then pull out of their bag of tricks when a particular feeling is called for, he called "illusions." I had to learn them, but in their lowly technical place. They got respect but not importance. They belonged to "cello history," not to me. Gustavus taught me how to find my way into the notes "on the right side of the door," as he used to say: at the point of feeling's entry, rather than its exit.

He did a good job. When I came back from Brussels, the critics—who suspect all prodigies of being mere technical wizards—were waiting for me. If you don't make them cry by playing "The Swan" from *Carnival of the Animals* as an encore, they try to drive you back to junior high school by criticizing your "musicality."

I made them cry, thanks to Gustavus. But not with the Saint-Saëns. I got them with an angular, dissonant Crumb piece, or a strange Kurt Weill movement, a jumpy Ives sonata, a transcription from one of Bartók's folk melodies. *Nobody* cries at that stuff, usually. But Gustavus was right: if I put feeling into playing every type of music, then people hear feeling coming out of every type of music.

There's something else I remember from that first meeting with Gustavus. He asked me a final question: "Why do you want the cello to play?"

Of course he phrased it this way because that was how it would have been constructed in German, with the verb at the end. But to me the structure seemed carefully devised, and struck a special kink into the question; it was different from "Why do you want to play the cello?"

I had looked at a lot of instruments in music stores around D.C., and snuck into the school band room for a few nights to mess with the things people left under their chairs there. I never came across a cello anywhere, but I tried everything from timpani to the accordion. None of the instruments was challenging enough. I mean, you could *hold* a violin, it was so small. A trumpet had only three lousy buttons to push. A bassoon required great lungs to move a nice-sized piece of air, but otherwise so what? The big instrument everybody was supposed to start with, the piano, was the worst of the bunch. All the notes were already *there,* all laid out for you simply to tap in the right sequence. You didn't have to discover anything or figure out your own way to get somewhere. It was like reading.

The instrument that almost won was the slide trombone, because none of the notes were marked, and I loved the sound of it. But I hate touching brass, and

I didn't think I was meant to blow air for my music. My music was in my hands, not my mouth—I just knew this.

Why do you want the cello to play? Here in the rustling of Iowa trees, I love remembering the woodsy Swiss way he rasped it. There's one other moment in my life when the sound of that sentence came back to me. I was going to play Britten's Cello Symphony with the Chicago Symphony Orchestra under Giulini, and there was this one tiny quick passage, just a couple of measures between bigger things, that I could not nail. Giulini pestered me through hours of rehearsal— he too could hear that although I was playing it decently, I was missing something that was within my reach; it dipped below the level of the rest of the performance.

The night of that concert came and I still didn't have the effect I wanted for those notes, but there was nothing more to do. We played the piece; it was going beautifully; then all of a sudden we were at those couple of nasty little measures and I found what I had been missing. I played them superbly for the first time. Giulini bugged his eyes out in this dashing, handsome frown, as if I had been holding out on him all week. But I hadn't. At the very minute I started the passage I remembered, for no reason, Gustavus's voice saying *Why do you want the cello to play?* and the tone and the cadence of the syllables and the number of beats were exactly right, and I just played his voice. It was eerie.

Okay, so what did I tell Gustavus? I finally just said: "I want the cello to play because it's big enough to hold me." He smiled very briefly, then nodded seriously. I suddenly pictured my body actually curling up inside one of the things. I felt for just a second the hard curved side piece against my back, saw the darkness, smelled wood and shellac three centuries old.

I took the bus home and told Taxi I was going to study the cello. He popped up and said he guessed we'd better go get me one, eh? We found an okay cello in a Silver Spring music store that specialized pretty much in rock-and-roll instruments. It seemed to have decent sound but a very difficult neck; when I sawed it in a charade of "trying it out" I didn't have any of the lucky rhapsody I'd found in Gustavus's room. I sounded wretched. Across the store, a black salesman in the electric bass section yapped: "Oh *yeah*! Like, *grunt*, Martha!"

We took the instrument. During my first practice session three days later, after I wobbled through a horrendous low A, I heard Taxi's voice from three rooms away in a shockingly perfect imitation of that jive clerk: "Yeah!" he said, "Grunt, Martha!" We howled together, three rooms apart, for ten minutes. The cello had a name.

I wriggle out of my sleeping bag and unfold into a standing position like a photographer's tripod; my arms and legs are ridiculously long and scrawny. Across the little clearing Taxi is silent in his sleeping bag, curled like a croissant with his back to the ashes of the fire he must have built.

Grunt Martha. I wanted a cello of my own, of course. But we didn't really know what to look for. So I played Martha for a couple of years, and to tell the truth, I think a few of my nifty little qualities of attack come in part from the necessity of having to deal from the beginning with that tough old neck.

Then one Monday, after a week in the mountains with Taxi, I went for my lesson and climbed the stairs to the back room of the house in Takoma Park and knocked on the door, all as usual, and when it was opened I said, "Good afternoon, Mr. Gustavus" as I always did, but instead of my old Swiss gentleman in loose pleated gray pants and a starched white shirt and dark-blue suspenders and a burgundy tie, there stood a naked college boy toweling off hair longer than mine. I tried not to look at his penis, hanging there in front of him like the soft unfinished beak of a bird I once cracked out of its egg on a school field trip. Behind him an equally wet girl with big tits was drying her legs. "Who is it?" she said. The boy nodded and passed the question on to me: "What's up, chickaleenie?" I stammered something and the boy

118

recognized Gustavus's name. "Oh, you're looking for the old dude. He dropped the body last Saturday. We moved in this morning." He grinned. Then he noticed my cello case for the first time. "Hey, was the old dude a musician? I'm a musician too!" He turned to tell his girlfriend that the old dude was a musician. I turned and bumped down the stairs.

I was already off the porch and on the front sidewalk when the landlady called me back. She acted like I knew what had happened and what she was doing when she took me in a weary way down to the basement, where in a corner near the steps a couple of small boxes stood on top of a cello case. "Most everything went to Switzerland after the cremation," she said. "There wasn't much. To send, I mean." She pointed to the pile. "He had a will in a drawer, not legal but what the hell. He paid his rent and I got no claim. He said these things were for you." She turned to go back upstairs.

I stopped her. How did he die? I asked. She shrugged. "It was quiet." A landlady's answer, I guess. But why had the boy upstairs been so surprised? Students must have been hauling cellos up those stairs all morning.

The landlady looked at me like I was crazy. "What students? You mean other people taking lessons? Ha, that's a laugh. Don't you know?" I didn't. "Well, you were the only one. His only student. He said he ran one ad in one music journal every issue for five years.

A few years ago some people came, but I guess he didn't take them. They never came back after the first time. No, you were his only one."

I thanked her, and carted the boxes and two cellos to the bus stop, making two trips. When I got home, Taxi was amazed at the news and the gifts (the boxes had my music compositions inside). He acted all sympathetic for a couple of hours until I finally said, "Look, he wasn't my grandfather or anything, okay? I'll find another teacher." Then he asked if I could tell the cello's quality by playing it; I told him I didn't want to play it yet.

So we found a little gnome in a shop in Foggy Bottom who identified the instrument right away as one of the sixty or so cellos made by Sergio Bianchi. This one was dated 1721. The guy seemed delighted with this, so Taxi asked if that was an especially good year for Bianchi. The gnome looked anguished and said, "Bianchi was not a chianti vineyard, sir!"

The big point was that a Bianchi cello was quite a prize. "A prize, that is, for a few of the more eccentrically gifted players," said the gnome. "It is rather feared by the mainstream. Its idiosyncratic properties reward only the hand of a perfectly matched player—but under such a hand it is astonishing. Its arrogance is justified, its darkness lucid, and its mendacity coherent." Kind of like a good chianti, maybe.

When I started playing the Bianchi, I recognized

120

some of its idiosyncrasies right away. And I recognized in them—or at least in what they seemed to require from me—a lot of Gustavus. The more I played, the more I realized that the quirks of my teacher and his cello built up a command of fundamentals often dismissed as marginal or strange, unique things that set me off from every other player I know.

Now that I'm thinking about the Bianchi, I want to see it, for some reason. So I walk over toward the bus instead of crawling back into my sleeping bag.

The bus. There it sits, its two tones of ugly blues looking black and silver in the moonlight, its silly face looking eager and fresh. What a funny beast; I almost expect its headlights to blink as it wakes up under my stare, and its wheels to start walking it my way for a nuzzle under the insignia. Cute. Not wise, necessarily—but cute.

I open the side door and step inside, sliding the door back a bit with the idea of leaving it about half closed. But—I swear—as I begin to let go of it the handle springs out of my hand, and the door covers the last two feet of gap with a determined little whoosh and click, and it's shut.

Oh well; maybe the turn signals are flashing alternately, too. Just you and me, bus.

Suddenly I am very snoozy, and my eye lands on the guitar and the duffels and my cello case, and I see that I can rig up a kind of mattress of the bags between the instruments, and sack out in comfort. I shove things around and in five seconds I'm cozy as the guitar and cello in *their* plush-lined cases.

My last thought is of the guitar. Why did Taxi want it? Tonight after dinner I was practicing a little and Taxi was out in the woods somewhere. I was tuning a string when I thought I heard a chord from way off in the distance. I stuck my head outside the bus and thought I heard another, this one sounding like a minor seventh. I checked; the guitar was gone. Was that Taxi out there? Did he even know what a minor seventh was?

13

I wake up early; those tiny damn wraparound windows let in a lot of light.

As I step out the first thing that hits me, even before the light, is the smell of smoked fish. Taxi, sure enough, is bent over a frying pan, holding a spatula and two tin plates with steaming triangles of corn bread on them. Taxi can certainly make do on the road. I sniff again: There's coffee too.

"What time," I growl.

Taxi shows me a quick smile, squints up at the sky even as he slides his spatula under the stuff in the frying pan, and says, "Oh, say about 6:10."

"What are we, Daniel Boone or something?" I snap. I walk two steps back to the bus, stick my head through the passenger window, and read the dashboard clock, which Taxi repaired before we left. "Hey, it's 6:12. Not bad."

"Call me Natty Bumppo."

"Who?"

"A great scout."

"Sounds more like a well-dressed clown. But I'm impressed. Where'd you learn to tell time by looking at the clouds for three quarters of a second?"

"It's the sun you look at, actually. I learned as a Cub Scout."

I fake an open-mouthed gawk, but it turns into a yawn. When it's over I say, "*You* were a Cub Scout? Those fruity blue uniforms with the pockets that button *open*? Getting into knots and woods and things in those dopey beanies?"

"Oh, I got myself into a lot of knots and woods without the help of the Cub Scouts. I still do so." He hands me a plate. I sit down and motion for some coffee, which he is already pouring. I take a bite of the smoked fish; it's very salty, maybe the best thing I've ever eaten for breakfast.

After we eat, he tells me that earlier this morning he found a clear stream running through a small copse of oaks beyond the pines; maybe I'd like to wash up? I grab a towel and some organic biodegradable guaranteed-not-to-get-your-hair-clean-but-it-makes-you-smell-like-a-Popsicle-and-doesn't-pollute-the-waterways shampoo, and find the stream easily. There's a spot where the water's about two feet deep, which is enough for me to splash around in for five minutes. Strangely, it reminds me of Europe—the many times

I've had to take cold-water baths in the morning even in the nicer hotels.

I stomp back to the clearing, shaking my hair dry. I'm one of those people who intentionally make a lot of noise in the woods, mostly to hide the fact that I couldn't be moccasin-silent if I tried; Taxi is one of those people the animals accept as an equal five minutes after he enters a forest. I stumble halfway into the clearing with my head down and my hair swinging, and pull up to make a joke about the shampoo. What I see stops the joke cold.

Taxi is sitting on a log away from the fire. He's holding the guitar. And he's holding it in a familiar way, his right hand gripping a pick and his left casually draped over the neck. I gape. All I can manage to say is:

"What's this?"

"A guitar," he says, looking down and pointing to the hole. "The music comes out here."

Before I can snap something back, he starts to play. I sit down quick, on the spot. It's startling: Taxi is *good*.

He begins with a series of chords, a simple progression he plays intelligently, making the most of the tune. He's warming up, extending the chord sequence into a long intro, playing it over and over, dropping into a rhythmic arpeggio here, seeing what feels good. I watch his hands. They're different hands from the ones that make birdhouses with tools; now they're a team of unlikely pals, one of them bouncy and unpredictable and good at dancing and jokes (the right hand, strumming), the other serious and lean and precise. I should have paid more attention to Taxi's hands all these years. They would have been nice to watch even making corn bread.

I look at Taxi's face, too, and I realize this is one of the few times I've been around when his face isn't either watching me or responding to something I say. It's a good face, thin but fresh looking. Maybe even a tiny bit dashing, with long diagonal hollows running from behind the cheekbones to the chin. If Taxi were big and had black hair, this feature would probably be that cruel-sexy thing so many girls and women look for in men they hope will break their hearts. In Taxi it's just a little spicy touch.

I'm looking at his face when he suddenly opens his mouth and starts to sing. I nearly stand up and run from the shock of it. Taxi singing! It's as if one of the mockingbirds that hang out on the roof beneath my window at home started reciting the Pledge of Allegiance. I'm so startled I miss the first couple of lines of the song, and when I settle down and try to catch up, I wonder if maybe they were some kind of key, because it's pretty weird stuff, about silver saxophones and cracked bells, with a refrain that says: *I want you*. For a while I really try to follow the lyrics, full of children in Chinese suits and broken politicians,

126

but by the last jangly chorus it's obvious there's nothing to get. Whoever wrote it was slinging some flashy ideas onto the chords to see if they stuck, like spaghetti slung against a wall in an old Italian kitchen to see if it's done. *So ba-a-a-ad,* like the song says.

Taxi finishes with a decent fading of the chord sequence, and finally stops. I clap a little too long. He doesn't look up except to say " 'I Want You.' Bob Dylan."

"The Christian guy? So is that some kind of biblical symbolism number?"

"He wasn't Christian yet."

"What was he?"

"Greenwich Village Jewish."

Then all of a sudden he's off into another song. This one starts with a raw, twangy line of notes played on a single string. Taxi knocks me back with a new voice, singing as if he were completely different from the breathy guy who was just sighing about lonesome organ grinders and fragile stuff like that. This time he pipes up brash and punky, telling the story of a guy who gets suckered by a girl and takes the job of driving her car instead of following his other prospects. The punch line: She's got no car—but now she's got a driver, and that's a start. Cute. Not wise, but cute. I laugh.

Taxi finishes it with a wild duet between voice and guitar, singing this perfect dopey phrase, "Beep beep uh beep beep yeah!" in gargly falsetto and answering it with sassy twang figures played on the high string.

This time I clap the right amount of time. He looks up with a slight smile. "The Beatles," he says. " 'Drive My Car.' "

"*That* was by the Beatles? The only stuff I've heard is nice ballad-schmaltz, just right for pops orchestras. This had some teeth."

He smiles again. "Oh, the Beatles could bite, all right. Especially in the beginning, before they forgot they were a rock-and-roll band."

"One of my teachers once told us, 'Without the Beatles no one from my generation would have lived past seventeen.' Is this what she meant? 'Beep beep yeah'?"

"It does probably look a little trivial, as a . . . cultural force. But songs like these put a lot of people into motion."

"You mean making them dance the Funky Chicken and the Swim?"

"No. I mean making them recognize that they had a lot of energy and could start doing things with it."

"Things like getting into the Chippewas?"

He doesn't answer, but starts to play again. This one's a jaunty tune with completely brainless lyrics and an uncomfortable premise: two people humping on a blanket under a boardwalk with people walking above (every time the chorus rolls around I plead that he won't rhyme "above" with "love," but of course every time he does). It's a ridiculous song, but I have

128

to admit Taxi sings it with a fantastic range, slipping into a bold, clear falsetto on the first and last syllables of the chorus line, "*UN*-der the boardwalk . . . down by the *SEA*-ee . . ."

I don't bother to clap, though I say, "Nice pipes, Tax."

"The Drifters," he mumbles. " 'Under the Boardwalk.' "

The next song is a little more interesting, kind of a jerky syncopated melody line with more pseudo-poetic lyrics, this time about picking up stitches and rabbits running in the ditches and beatniks out to make it rich. I know what beatniks are, from reading *On the Road* in the tenth grade. It doesn't surprise me that they were trying to get rich by the late Sixties.

When it's over, I make the kindest comment I can: "I can't decide if it's interesting, or just crummy poetry."

Taxi gives an understanding nod while flexing his left hand. "Yeah, well, Donovan was usually such a dipshit—most of his other songs are about sages and mages and purple-velvet-clad ladies of the sun. This one's kind of a relief because it has balls."

I've never heard Taxi talk like this, and he seems suddenly to realize it. He looks up and smiles with a blush. "Slipping back myself," he says. "Should I say, 'Donovan was usually a lyricist of meek and inconsequential sentimentality, but this one has a more aggressive cutting edge'?"

129

"I prefer 'dipshit' and 'balls.' "

He pauses with his hands above the strings then rolls into a sort of fake Baroque arpeggio and a burly vocal whose third-grade rhymes ("liar/higher/fire") don't quite ruin a haunting kind of urgency, especially in the command *Come on baby, light my fire*. Without pausing long enough for me to tell him I liked that one, he jumps right into an inane tune about respect (which rhymes "honey" with "money" twice), then a vapid tune about doing something called *grooving*. Last, he plays a perverted samba with yet another weird lyric, one line of which will probably keep me awake in the middle of the night someday, with its *bom-da-bom-da-bom: bom-bom* rhythm and obscurity:

> *Won't you come to my house and black cat bone?*
> *Take my baby away from home. . . .*

"Thanks, Tax," I say when it's over. "You're pretty good."

He shrugs. "I haven't played these songs for a while, especially that Bo Diddley."

"Who taught you? Doesn't really sound like a classical background, but you're pretty precise inside the musical structures."

He laughs and puts the Martin gingerly back into its case. "The radio."

"Lessons over the air?"

"No, no. I listened to songs on the radio and picked

out the chords by ear. That's how everybody learned to play back then. Back then almost everybody played guitar, and played these songs and others like them."

"Am I going to hear the others?"

"Some of them, if you can stand it." He smiles and stretches, looking at the clouds over the trees. "You were patient. Of course, you'd really only get a true picture of what the Sixties music was trying to do from hearing the records, but. . . ."

"It was pretty stupid stuff, but I liked hearing *you*, Taxi. You're good. I mean it. You ought to play some *music*."

He snaps a look at me to see how nasty that comment was, and sees that it wasn't really nasty at all. "Yeah," he says, "well." He looks up again. "It's going to rain. We'd better hit the road."

An hour later we're two hours from Des Moines, and I ask: "Did 'everybody' really play the guitar back then?"

He smiles. "It sometimes seemed that way. The average two-person dorm room contained 1.65 guitars. Men played them, women played them, antiwar activists and lesbian rights crusaders and Jesus freaks played them—they were everywhere."

"And did my mother play the guitar? I mean, If 'everybody' did . . ."

Taxi stares ahead for a minute. Then he says, "No. No, she didn't play the guitar."

"What instrument *did* she play?"

He sighs. "Well . . . she had an autoharp."

"A what?"

"An autoharp. It's kind of a pegboard with strings and . . ."

"Oh jeez. Not that thing the dippy music lady used to bring around for us to play in the first grade when they wanted to make everyone feel like a musician. 'Just press these little knobs, boys and girls, the notes are clearly labeled, and you get real music! Now—one, two, one, two, C C C, G G G, hum hum hum . . .' Yuck de la yuck. I remember her telling us that it was impossible to make a bad sound on the autoharp—you simply couldn't hit a wrong note."

Taxi flashes a wry smile.

"You say she 'had' an autoharp. Does that mean she didn't even play it?"

"She didn't play it much. She kind of got it out once in a while when . . . well, once in a while."

"When what?"

"Sometimes a few friends would come over to make some music. Play the guitar, maybe some harmonica, fool around."

"*She* played the harmonica? I *like* the harmonica," I say, unsure whether I want her to play an instrument I like.

"No," says Taxi, almost apologetically, "she didn't play the harmonica."

"Who did?"

"I did. Not very well, I'm afraid."

132

I grunt. "So a couple of your friends would come over—and I bet they were your friends more than hers—and you guys would say, 'Hey, let's play *Beep Beep Yeah*' and my mom would get out her autoharp and join in?"

"Kind of. But she didn't really join in. And not right away. Usually she got her autoharp out after we had played a couple of songs."

"But then she didn't join in?"

"Well . . . she would wait until we were between songs and she might, you know, start strumming it a little, very softly, off in another part of the room, almost as if she didn't want anyone to hear."

"Uh-huh. 'Almost.' Just a complete coincidence that she happened to haul it out when you guys were between songs, right? Her urge to make a little harmonious self-fulfillment had nothing to do with the boys rocking away across the room."

"She didn't play the guitar, and this was a way . . ."

"Of getting some attention. Why didn't she go and get a guitar like everyone else and teach herself how to play the three chords that are the basis of every rock-and-roll song I've ever heard? If everyone else could do it, why couldn't she?"

"To tell the truth, she didn't like the guitar."

"Because the notes weren't written on little knobs?"

He laughs. "No. Because . . . she thought the guitar was an elitist instrument."

"Elitist? You mean like tennis?"

"No." He puzzles about how to explain. "What she meant was that the guitar encouraged people to elevate themselves above others by dint of their skills. The poorer player would feel awful and the better player superior, a whole system of unjust discrimination designed ultimately to keep people apart."

I'm staring at him in disbelief. I stare for maybe a full minute. "She *thought* that twerpy crap? Did you laugh at her?"

He looks straight ahead. "I had just married her."

"So—there you were sitting, having fun, maybe making some decent music, passing around a jug of wine, nodding and smiling, when all of a sudden into the nice atmosphere creeps these stinky twinky sounds. Nobody looks, but everybody knows. Ms. Anti-elitism has whipped out her equalizer over in the corner, 'Please, guys, don't pay any attention to me, I'm just fulfilling myself, C C C, G G G . . .' "

"Lighten up, okay?"

"It's hard to resist."

"Believe me, I know."

I study him. "Did you ever cut loose?"

"Yes. I came East with you sixteen years ago."

"No, I don't mean that. I mean . . ." But I guess he's answered the question. Coming East with your daughter may not have the momentary zip of a few sarcastic lines, but as statements go, it packs a definite punch. Even if the idea wasn't yours at first.

We ride on for a while without talking. It takes

some time for the sound of that autoharp to die down for both of us, though for me it's imaginary and for Taxi it's memory.

But the subject of music and my mother cannot be dismissed until I have asked the single question I must know about everyone. So, somewhere in between Davenport and West Branch, I ask: "What kind of music did my mother like?"

Taxi doesn't respond right away. He has been musing, and he appears to continue musing. Just when I am about to repeat my question, he starts to tell me a story.

"During the summer of 1968 there was a nation-wide talent show, a contest, on television. We were living in a house with a friend of ours, a guy named Christy. Christy's grandparents had left him this nice Victorian house the year before, and he had invited us to spend the summer there in return for my helping him with some basic repairs.

"Christy watched a lot of television at night. He also smoked a lot of dope, which is why he could tolerate the TV; he was one of those people who thought everything was fascinating when he was stoned. There were kind of two living rooms, connected by a wide archway; really it was like one room with a small break in between. One room held the stereo—Christy had quite a good one—and the other the TV. We couldn't have the sound of both things on at once, and I didn't like TV except for baseball games, so Christy—and your mother, who used to do her macramé or string beads in front of the television—let me play music in one room while they watched without the sound on in the other. Christy said he liked it better that way—watching a cop show with a Tim Buckley sound track was the height of comedy. Your mother liked mostly to watch quiz shows. She was studying the ways people compete, for a book she was thinking about writing on noncompetitive games.

"Whenever this talent show came on, though, Christy would turn the sound up and concentrate. He ate it up. So did your mother, though for different reasons. Christy was 'goofing' on the show. Your mother just seemed to like it.

"So the talent show bits became a kind of thread running through the summer, a stronger thread as the thing rolled on and the acts became familiar.

"The contest was structured so that on a particular night in August the two acts that had won every previous match would square off, head to head, in a live broadcast before the judges. It was talent against talent for the grand prize—a recording contract, a lot of publicity, a big chunk of money. There was a lot on the line, and it was pretty exciting for people who had followed the thing all the way, especially if one of the final two happened to be a favorite.

"I had missed quite a few of the broadcasts; your mother and Christy had missed almost none. I decided to watch the final with them, but I was expecting to be bored. Christy was cleaning some pot in a shoe box lid in the corner, very stoned, half watching, half joking about whatever came into his head vaguely related to the show. Your mother was intent. I asked who they were rooting for. Your mom just shook her head at the question. Christy said, 'Bobby Kennedy.'

"The first act came on, a sweet, pretty girl named Melanie. That was all she went by—no last name—just Melanie. She wore a baggy shift dress and played a huge acoustic guitar; you got this feeling she chose it because it made her look tiny and adorable. She shook her long hair out of her face as she played a few earnest chords. Then she started to sing in a wispy high voice.

"Her first song was about trying not to step on insects and flowers, because they were alive too. Her second song was about stopping to smell flowers even in the rain, instead of hurrying inside to get dry, because if rain didn't hurt flowers it wouldn't hurt you. Her third song was about kissing by candlelight *in* the rain *near* some flowers.

"She was a big hit. She giggled when people applauded her at the end of her set. She bent down and sniffed the flowers some people threw on stage. She dimpled and said, 'I love you, love is all we need,' which was not an original line but it worked. She was a smash. I hated her guts.

"I asked your mother what she thought very sardonically. Too late I noticed a teardrop oozing down her cheek; she said, 'She's beautiful. She's a china flower on a crystal love stem.'

"I was pretty shocked. There was a kind of challenge in her voice, a response no doubt to my mocking tone. Dazed, I watched a couple of cigarette commercials whiz by, and missed the title of the final act, catching only the word 'Family' in it. Shit, I thought; here comes a grown-up Melanie who sings *with her kids!*

"But I was wrong.

"A dissonant chord of horns and guitars and voices *shrieked* out at a terrific volume and hung in the air. As it ended, the stage was full of bouncing people dressed like Cherokees and jesters and satin strippers, hopping in unison and singing: *bumbum-bum-bum-bumbum . . . bumbum-bum-bum-bumbum . . .* Then all together they broke for their instruments, and three of them took the microphones and everyone blasted as they sang *DAAANCE TO THE MUUUSIK.* I was on my feet howling before I knew it.

"Sly and the Family Stone. All cousins and sisters and brothers, except the white drummer, who was probably related somehow. All of them scary and wonderful in their swinging fringe and bobbing Afros and stomping moccasins. They never let up, growling and bouncing through complicated rhythms, screeching out trumpet solos—the trumpeter was a girl in a white Afro wig—buzzing bass lines, dropping into a capella. At the end, Sly, who was wearing platform boots that slithered up above his knees, leaped at least four feet into the air, hung there for what seemed like several seconds, and came down with those boots on the final beat, crushing to dust the China Flower on the Crystal Love Stem. Melanie was destroyed.

"But I was transformed. I yipped and spun and stomped. I caught the lyric reprise of every song— 'Dance to the Music,' 'Sing a Simple Song,' 'Want to Take You Higher'—and yowled along with Sly. I picked up a T square lying in the bay window we had worked on that day and played along with every raunchy wah-wah guitar solo.

"Christy was chuckling from the corner, saying 'Sing it, brother' and 'Ladies and gentlemen, the hardest-working man in show business!' But during one especially zealous spin, when I was whirling in a 360 with my guitar held over my head, I happened to zip my eyes over your mother's face.

"She was staring at the set with tight white lips and a fierce hatred I could never have imagined in her; fortunately, I thought, it was directed at Sly and the Family Stone.

"Sly's final song came crashing toward an obvious close. I dropped the T square and whipped into my finest James Brown foot-sliding reverse-twist cross-stage shuffle move, which I had watched him do thirty times a night twice a year at the Howard Theater downtown, and had practiced for hours in front of a full-length mirror in my bedroom. I was wearing sandals, and they weren't ideal . . . but all the same, I would have been okay if at the height of my move a quick little beaded moccasin hadn't snapped out and kicked me deftly just above the left knee. I sprawled wildly, snagged the cord to the television, wrapped it around an ankle, and with my other foot flipped the T square up as I went down. I hit the floor and the steel T tomahawked the screen and exploded it with a hissing pop into a thousand little glass nuggets. The talent show was over.

"I wasn't hurt, at least not by the fall. I hoisted myself up on my elbows and stared at your mother. She was watching the TV as if it had never blown up, and silently stringing her beads. From the corner came Christy's voice: 'Bummer. Now we'll never know who won.'

"When he said that, your mother met my eyes. 'I think we know,' she said. 'I think we know who *really* won.' "

Whew. That's the longest Taxi has ever talked to me without asking for a comment or letting me hedge a question in. I wait a couple of seconds to let him catch his breath.

"So," I say, "my mom liked stupid girl sweetsie-flower-song singers? That was her favorite music?"

"That's it," he says. "That's really all she liked."

"And you didn't."

"Right."

"You *hated* it."

He nods.

I think for a minute. "And even though it's representative of one side of the Age of Aquarius, I bet you're not going to play me any of it."

"You're right," he says. "I'd trade the Martin for a copy of *Sly's Greatest Hits* first."

After a while I go in the back and practice.

14

We're about twenty minutes from Des Moines when I put the Bianchi away and scoot up into the front seat to eat a sandwich. "I wonder what Des Moines is like," I say, though I don't really. I usually have to say something after I stop practicing to let Taxi know it's okay to talk; he has this idea I might be practicing in my head, or strung out in my musical trance, or something. He'll stay quiet for a couple of hours unless I break the silence.

"We'll find out in half an hour," he says.

I gape at him. "We're going *into* Des Moines?"

He laughs. "You'd think it was 7th and T at 2 A.M., the way you say it."

I shudder.

"Dvořák loved it," he offers.

"Great," I say. "That's cool. Dvořák really knew how to party. What are we stopping here for, anyway?"

"I want you to meet someone. An old college friend named Gwendolyn. She was a very important person back in the Sixties."

"Important to you?"

"Also important to America. She was one of the founders of the first national feminist organization, or at least the first well-organized one. It got legislators elected, legislators defeated, laws passed, laws repealed, taxes changed, books republished, prisons revamped—and in the process it managed to convince a lot of people that our society operated under ridiculously limited ideas of what women were."

"Consider my consciousness raised."

He smiles a private smile, looks back at the road, checks a sign. "We turn off at the next exit."

"How do you know where she lives?"

"The same way I know where your mother lives. I send her the newsletter and I get change-of-address forms. She's been in Des Moines for two years."

"I wonder why."

"So do I."

Taxi needs to stop a couple of times to get directions. We drive through a lot of neighborhoods that get increasingly shabby. Finally we turn off one kind of busy street with a few massage parlors and Vietnamese markets and bars into a dark street overhung with some trees that drop berries. One falls onto the roof with a *pong,* a perfect F sharp, kind of like a warning. Taxi finds the house and parks the bus.

"Was this woman beautiful?" I ask.

We're squashing berries underfoot, so we slow down to pick our steps more carefully. "Yes. Ironically, it was one of the things that made her so successful. Nobody could dismiss her as a radical kook. When she talked about men and women, people listened and believed it was coming straight from nature."

"Did you have a fling with her?"

"None of your business."

"Did my mother like her?"

"No."

We're getting close to some cement steps that lead up a small hill to a smaller yard.

"I'm betting it was tits."

He snaps a look at me and blushes. "What?"

"Tits. I'm betting this woman had big apples, and my mom didn't, and she hated her guts."

He shakes his head and his blush fades. "Your mother had found her own system of feminism," he says as we climb the steps. "She pieced it together out of some of her . . . familiarity with other cultures. She didn't like organizations; she wasn't a 'joiner.' "

By then we are on the wooden front porch. Taxi takes a deep breath and knocks on the door. " Gwen. Wow. We were such good friends," he says wistfully, as if we were leaving instead of arriving.

The door opens. A dark woman of about thirty,

with stylish short hair and mean eyes, takes her cigarette out of her mouth and says, "What is it?"

Taxi says, "I'm a friend of Gwendolyn Pierce's. My daughter and I are driving across country, and I wanted to stop in to say hello. I haven't seen her for seventeen years."

The woman drags on the cigarette and nods, not in acceptance of what he says, but as if she knows true dark secrets behind our presence.

"We also want to sell her some heroin," I say, "and a subscription to *Playboy*. A *lifetime* subscription."

Taxi kind of gasps but the dark woman takes it well, squeezing one eye my way while she closes the other behind her smoke. She smiles a crooked female-to-female smile at me and says, "Why not?" Then she turns back into the house, leaving the door open. We follow. I noticed that her corduroy pants have cuffs and she's wearing wing tips.

We walk straight through the house into a kitchen, where another woman is doing something at the sink. Her back is to us; she's very curvy, wearing a very old, thin blue-jean skirt that clings to her hips, and a tank top. Her hair is red. There's country music on the radio, playing pretty low. The dark woman watches the other woman for a second, then says, "I believe you know this person. Watch out for the young one; she's got somebody's number." Then she goes back down the hall.

The woman at the sink turns around; when she sees

145

Taxi she puts her wet hands up to her mouth. They're beautiful hands, but not as beautiful as the mouth. I can tell this woman doesn't have *anything* that isn't beautiful. Right away, I feel her power. She's so powerful I can't believe she's here, in a room, in a city; I can't believe she's anywhere. It's like she should be protected.

I can't help staring, but it's okay—she's just looking at Taxi and doesn't even know I'm there. I can see from the disbelief and joy in her face that she and Taxi were good friends, like he said; I can see so much feeling in her eyes, it kind of shocks me. I feel a little jealous—but not of her. Of Taxi. I want this woman to notice *me*. What's even weirder is I feel like a little girl, a nine-year-old, and it's okay—in fact, it's terrific: that's how I feel I *want* to be, around her.

They step toward each other and very slowly, with their eyes and small, tender smiles meeting the whole way, they take each other into a hug. It's too much to see. I walk out quietly, and head down the hall.

"Hey," a voice says from a room as I pass. I look in, and see the dark woman sitting in a chair reading a book. "I was counting on you to chaperone."

"Sorry."

"Are they tearing each other's clothes off? Making happy animal noises? Clinging lip to lip?" She smiles, but I can see she's nervous.

"When I left everything was still fastened," I say coolly.

She doesn't say anything. Then she sighs, puts her book down, and stands up. "Guess I'll go," she says loudly, and waits. Nobody says anything, and she nods, looks at me again, and walks right past me and on out the front door, which she closes with a slam.

I watch the door for a second and turn around. Taxi and the woman are peeking around the kitchen doorway. He's looking confused, and still a little flushed from the hugging; she looks tired. But still beautiful—definitely beautiful.

"Where's Dolores?" she asks me.

"Her wing tips took flight."

Taxi laughs. Gwendolyn looks at him and says, "It's not funny."

"I wasn't . . ."

"This is great for you," she flares, "a quick visit, a big rush of nostalgia, a quick feel, then back on the highway." She sighs, and looks around. "Jesus. *I* live here." She sounds like she can't believe it any more than I do.

"Gwen, I . . ."

"Forget it," she says, waving a hand gracefully. "It's not your fault. I'm sorry." She puts a hand up to his face, but I can tell that's as far as the touching is going to go from now on. "It is nice to see you again. We were such *good* friends."

I can see Taxi frown as he deals with the past tense of that, and I see why he said it on the porch. The past tense makes a better opening line than a closing one; it leaves you more room.

But Gwen is walking down the hall toward me, with her hand out.

"Are you Cabot's daughter?" She smiles.

I nod.

"What's your name?"

"Sib," I manage to say. "Sibilance."

She takes it without a flicker. "Good name," she says. She turns to look at Taxi over her shoulder. "Good name," she tells him, too.

"I agree," he says. "But I get no credit for it."

"Your wife?" she says to him. I snort; she turns back to me.

"No," Taxi says, walking down the hall. "She named herself when she was eight. She came up with Sibilance." He smiles at me proudly; it's very goony, so I laugh, but I know I'm blushing too.

We go and sit down in another room, a poorly lit living room with a toast-colored couch that's getting old and some modern leather chairs that seem peculiar in the dark. She offers us tea and stuff and we don't take it.

"Okay," she says, "why have you come?" She looks from Taxi to me and back to him.

"I . . . well, now that we're here I don't really know how to explain, Gwen." He looks genuinely perplexed. "I don't know what I expected to *do*."

She smiles softly at him, but briefly. She's not going to help him. I probably should, but I wait.

He's frowning. "I don't suppose I really expected you to *explain* anything. I guess I wanted to introduce you to Sib as a kind of *example*. Of . . . of someone who . . . whose life . . . is a reflection of the ideas we held back when we . . . knew each other first." It doesn't sound good to him. He adds: "As you are now. Not as an artifact or anything."

She looks over at me. "Why is he doing this?" she asks.

"He's taking me to meet my mother for the first time," I say, without thinking about whether or not I should tell her. "He wants me to understand the stuff that was going on back then, so I won't feel too bad about why my mother unloaded me."

She nods seriously and looks back at Taxi, raising her eyebrows in a question. "Was it Connie?"

He nods.

"You were married?"

"Yes," he says. She nods again, and looks back at me.

"There's nothing you can understand about your mother by meeting me. I'm sorry." She says it sincerely, as if she really wanted to help.

Taxi starts to explain what he was looking for, but I cut him off: "Why?"

Gwen stares at me. I repeat: "Why do you say that?"

She looks across the room and thinks, then back at me. "Your mother and I had nothing in common except your sex. And that's very little, really."

Taxi almost gasps; she turns on him with a thin smile. "Not exactly the party line from the old days, is it?" she says. He shakes his head. She asks him:

"Why did she leave you?"

"Well . . . *I* left, actually," he says. "We were in San Francisco, and Sib and I went east together." He says it like I was a helpful companion, not a one-day-old baby.

"Why did she throw you out, then?"

"Damn it," he says, "why do you assume that I had to be 'thrown out,' as if I had no initiative? . . ."

"You wouldn't leave a tree house in a lightning storm if you thought the tree liked you," she says. "And you're loyal. You don't break commitments. So tell me why."

He watches her, then looks over at me. "She wanted to pursue some . . . pursuits," he says.

We all laugh at his loss for words. Then Gwen asks seriously, "Was it a bad labor?"

"That's not it," Taxi says. "It came pretty swiftly with the baby, but it wasn't so . . . superficial." He shrugs. "It just didn't take her long to see that her life wouldn't be all hers for a while."

"Was it 'all hers' before the baby?" She puts her hand on Taxi's. "Sounds like you weren't taking much from the marriage, my friend."

150

Taxi looks at her, and I see water bulging in his eyes. "I didn't think marriage was for 'taking,' " he says almost angrily.

After a moment Gwen turns to me. "Maybe I was wrong," she says. "Maybe your mother and I have more in common, more even than your father thought."

Our eyes are locked and I know what it is.

"You've got a kid," I say. "You left it."

She doesn't even nod, but I know I've hit it. We stare. Finally she says, "A boy. Not as old as you. He's eleven."

Taxi's sort of sputtering. He's half delighted that she has a kid and half confused: it's not exactly babes in toyland around this house. "Where is he?" he asks.

"Des Moines," Gwen says, like it was a thousand miles away.

"Does he live . . . with his father?"

"Yes."

Taxi waits two seconds, then asks, "Why, Gwen?"

She looks at him as if he's stupid. She looks across at me; it's time to help him out now.

"Oh, come on, Tax," I say. He snaps his eyes over at me. I look at Gwen and take a big breath. "I bet you're going to *say* it's because you're dykes."

She winces at the word and her eyes flare, and for a second I wish I'd played it soft and nice, but what the hell. This sister-bond thing can only go so far.

Taxi reels for a second, but immediately focuses in

on the key details. "Even so—you *could* have kept him. . . . I mean, it's an unusual, um, feature, but surely . . ."

"The courts might not have seen it that way. I didn't want it to come to that." She stares at him hard, daring him to quibble.

But he won't back down, not on the matter of a kid. "That's not really the reason, Gwen," he says softly. "That's not it at all."

"Oh no?" She glares.

He shakes his head. "You used to *love* to get cases like this into court. You said it was the quickest way to turn principles into law. You're not afraid of *that*."

"Times change."

"Principles don't. Not yours."

She stares at him hard, but the fire dims. Finally she exhales loudly, as if she's been holding her breath since her son came up. "No," she says.

I wait for them to say more, but they just look at each other mournfully. I can't stand it. "Then *why*?" I nearly yell.

She looks only at Taxi. "I found something I needed, more than another person I'd have to make into a man. His father could do that, probably better. I needed someone to make me into a woman."

"I guess it *was* pretty thoughtless of the little guy to be born a male," I say. "But I bet you would have found some reason to give away a girl, too."

She turns to me with her chin high, a flash of pride

152

showing through some terrible pain. "I love my child," she says. "But I had to learn to love myself first. You may never need to do that—you may be beyond it already. If so, you're lucky." She pauses, collecting herself. Then she just repeats: *"I love my child."*

"What's that got to do with me?"

"More than you think," she says. "Maybe you'll find out when you get to San Francisco."

There's a clumping on the front porch. The front door opens. The dark woman steps inside and stops briefly at the entry to the living room. Then she clumps upstairs.

Gwen gets up. Taxi and I don't. She looks down at him, probably debating whether or not to touch him again. She decides against it. "I like the newsletter," she says.

He nods and looks away. She blows me a kiss, then steps quietly upstairs. Taxi's either dazed or thinking too hard, so I get him up and out, and we walk on the berries back to the bus and head west.

15

We've been out back on the road for hours, sitting and watching the soybeans thrive. I ask: "Was it that big a deal, Tax? Having a kid? I mean, the only two Aquarian moms I've heard about both left their kids. Is it really impossible to pursue pursuits with a kid? *You* didn't find it impossible. *You* didn't give up everything, did you?"

He looks at me with a half smile, then back at the road. "It's a big deal, Sib. It's the biggest deal there is."

"Oh, come on. How long does it take to change a diaper? Twenty seconds? How many times a day do you do it? Six? That's two minutes. How long to give a baby a plate of your dinner mashed up—an extra fifteen seconds? Three times a day, that's almost another minute. I'm exaggerating a little, but, I mean, what are we talking about here? A couple of hours a day?"

He thinks, and says, "Why don't you get all A's in school? Why don't you have more friends? Why don't you join the Glee Club, or try out for cheerleader, or get the lead in the junior class play? I mean, you've got the talent and the IQ. What's the problem?"

"You don't win the Janáček Prize and the Rome Medal by spending your afternoons eating potato chips at Susie's house and your nights memorizing algebra equations," I snap.

He makes a face of mock confusion. "But winning those competitions takes so little time! I mean, how long are you on stage? A half hour for six or seven days? That's three or four hours. A final hour recital? Five hours. Once every summer, five hours! School isn't even in session—you shouldn't lose any time for friends and study and clubs at all!"

"Okay, okay. Ha ha, very clever. But I still don't see it. Where do the energy and time go? What do they *produce*? When I practice the second Beethoven sonata, I develop the ability to play the second Beethoven sonata. If you spend so much time with your baby, what do you develop?"

"I should say, 'Your baby's ability to play the second Beethoven sonata,' but that's not entirely true. . . ."

"You bet it's not. I learned that sucker all by myself. You were in the other room telling people about how soapsuds in the rivers kill snapping turtles, or whatever, and trying to get their eighty bucks for another

155

year's subscription. I was alone, and I had a problem, and I solved it. I'm not trying to say I don't think you're a good dad, Tax, as far as that goes—I'm just pointing out that I do *my* things myself. It's probably hard for you to admit, I guess, it's probably not what you think you're going to get when you do whatever it is that makes being a parent such a big deal. . . ."

"I beg your pardon. It's *exactly* what you hope to get." He laughs. "As far as that goes."

"Why don't you give me some specifics," I say.

"Specifics?"

"Yes. What are some of these things you had to do that took away so much of your life that you couldn't have macraméd an owl if you'd wanted to?"

"Don't put words in my mouth, Sib—I haven't ever said you 'took away' from my life."

"You said I would have taken away from my mother's."

He takes a deep breath. "Your mother would have discovered, as I have, that you add a lot more than you 'take away.' "

"Okay, okay. Very nice. Poor Mom, she missed out, her loss. But still, I want to hear what it was like. What does a baby do? What do you do with a baby?"

To my surprise, when he starts talking it's not sentimental stuff. But he looks sentimental about it. "The most noticeable thing at first, probably, is what happens to your attention." He raises his eyebrows. "It's suddenly got an object, and if you want to give any

156

attention to the things you used to think about, you have to think about them on a second level. The way you would, say, eat a sandwich while driving. The driving has your first level of attention, or it should; the sandwich gets what notice you can give it in quick bites."

"That's a basic thing about being a parent?"

He nods.

"Okay, I can see that. So, what about . . . me? Was I . . . would you say I demanded more than normal? Less?"

He looks over at me; I'm staring out the window. He shrugs. "I have no way of comparing. I'd say you were normal, I suppose."

"So how many hours a day did you have to be so attentive? Ten? Eight? Fourteen?"

He looks at me. "Pretty much all the time, Sib."

I frown and look doubtful. "*All* the time? You mean twenty-four hours a day?"

"Of course." He laughs. "I didn't have to sit and play dolls with you for twenty-four hours. But I was *there* for twenty-four hours. Babies need you when they need you. For lots of things."

"But you were starting the newsletter. When did you work?"

"I worked a lot over the telephone, quick calls. For interviews I usually took you with me in a little pack that hung inside my overcoat. I snuck you into a couple of libraries, too, when I had to do research. I went late at night when you were usually asleep. I rigged up a thing that looked like a fat briefcase, a kind of portable cradle. Used it when you were very tiny. You loved it." He laughs. "A librarian caught me once. Popped into the room while I was feeding you, sitting there pretending to read a book on the table, with my other hand down low holding a bottle that disappeared into the mouth of the case. She walked by and didn't say anything, but when I left that night, she checked out my books and said, 'How old is your briefcase?' I said, 'Six weeks.' She said, 'I have a spacious backpack that I take into a lot of movies. My backpack is eleven weeks old. Does your briefcase sleep through the night?' I told her it didn't, not yet; what about her backpack? 'Not yet,' she said, 'but it sleeps through movies.'"

"Wait. You mean you used to interview your people with me inside your coat? Congressmen, all that? With me right there?"

He nods. "You were pretty well tucked away, and usually pretty happy. You didn't cry much. I usually took you for a long walk in the streets first or to a museum, to show you a good bunch of things, so you wouldn't feel too bored during the conversation."

"But what did the *congressmen* say? Didn't they think you were a weirdo, doing business with a six-week-old kid using your tie for a hammock? I mean . . . I'd think you were crazy if you came to me like that. 'Congressman Butterface, how is the regulation on acidic phosphate measurement in secondary estuaries going in your subcommittee? And say hello to sweet widdle Esalen here.' Jeez. What a nut."

He smiles. "Congressmen meet a lot of nuts. A lot of them are carrying something more dangerous than a baby girl. And anyway, most of them were prepared to think of somebody writing a newsletter on environmental regulations as half a hippie, and thus strangely humanistic and unpredictable. But they were professionals, and the environment was a hot issue, and they wanted to get their names in print on one side or the other as often as they could, so they talked over your head." He shrugs. "What could they do? What could I do?"

I shake my head. "When did you *write* the newsletter?"

"At night."

"But you told the librarian I didn't sleep through the night."

"For the first eleven months you woke up a couple of times a night—three times, two times. I usually wrote between your first sleep and your first waking, then fed you, changed your diaper, sang you back to sleep, and wrote another hour or so. Then I went to bed too."

"Wait a minute—'*sang*' me to sleep?"

He laughs. "You were a music lover from the start, if you can call my renditions of Brahms's lullaby and the slow songs from half a dozen Broadway shows 'music.' After I fed you and changed your pajamas—"

"Why my pajamas? Why not just my diaper?"

"You usually threw up a little at every feeding. It's nothing—all babies do it. You kept most of it down; just the last bit would get caught on a bubble."

"I puked every night?"

He nods. "More like two or three times a night. But I tell you, it was nothing. You weren't sick."

No, but you must have been, I think. I hate puke. But I guess Taxi doesn't mind it. "Okay, back to taking me everywhere. What about baby-sitters?"

He shakes his head. "Couldn't find any that were good enough."

"What do you mean, good enough? They couldn't change a diaper?"

"Of course they could do that. But when I tried sitters, I asked them to talk to you a lot, to explain things they did in the course of the evening, and to do it in normal, simple language—no 'baby talk.' Most of them said 'Okay' to humor me, but I could tell they weren't going to do it. A couple even said, 'That child can't understand if I explain that I'm peeling an apple skin with a knife—why waste my breath?' I paid them and let them go on the spot. There was the same kind of resistance to reading to you."

"Reading? How old was I now?"

"I started reading books to you and showing you magazines when you were a couple of days old. You loved them." I snort, but he gets serious. "It's true!

160

And why shouldn't you like seeing bright photographs in an article about Kenya game preserves in *National Geographic*; or close-ups of butterflies in *Scientific American*? You had eyes, and as much intelligence as anyone in the world. Babies are born smart."

I look at him. His eyes flash; he's really charging. "So the sitters wouldn't talk to me or read me books."

"And they wouldn't show you your cards, either."

"What cards?"

"I made these cards for you. Kind of like flash cards. I bought some busted old books with lots of pictures of things—a couple of nature encyclopedias, a big book on trains, a dictionary of music with pictures in the back of all the symphony's instruments. I cut out pictures and pasted them on shirt cardboards and showed them to you a few times a day very fast, telling you the name of the thing on the card as you saw it. Blue jay, meadowlark, indigo bunting, yellow-shafted flicker, crow, bam bam bam bam bam. You loved your cards."

"Taxi, you sound like a nut. The sitters must have thought you were crazy. What made you think of doing all that junk?"

"You did."

"Oh sure. One day I said, 'Hey Pops—how's about a fleeting glimpse of a yellow-shafted flicker?' "

"No. You made me think of them by being bored."

"Bored? At two days?"

"Bored, absolutely, at two days. You needed things

to see and think about, or you got bored. It was very clear. It started our first night alone together, driving from San Francisco. We were in Nevada. When the car stopped, you got upset, and food wasn't what you wanted, or cuddling, or bouncing or burping or a new diaper. I sat there in a motel, holding you as you wailed, and I thought, as kind of a comic line, *Well, I can't blame her—Nevada's a pretty dull place*. So I went and bought a magazine, and held you in my lap, and went through the whole thing. You got quiet on the first page, and stayed quiet all the way through, and then stayed happy for half an hour. Then we did it again.

"I started looking for ideas for things you would find interesting. I made a lot of little wooden gadgets for you, things that kept moving once you put them into motion, visible gears that meshed and turned, little chutes that marbles rolled down, that kind of thing. I played the guitar for you a little, too. And I played records—every kind of music there was. We danced a lot. I built little cloth crawlways and caves in the apartment, so you could move around and explore."

I think for a while about all of this. "Well, my life must have been quite the whirl. But one thing seems pretty obvious, Tax. With all this cutting and pasting and vigilance, you didn't have time for any love life."

He jerks a look at me. "What makes you think that?"

"Well, it's obvious. I mean, I've never seen you with a woman. You've never mentioned a woman to me. Never said, *Hey, Sib, your dinner's in the oven, I'm going out, don't wait up*. You've either been working or messing around the house or waiting for me to finish a lesson or sitting in a concert."

"Do you think the only things in my life are the things you know about?"

"Yes. And I don't know about a love life, so I don't believe you've had one. Have you?" I look over at him. After a second he looks over at me and winks.

"You *have*?" I say, incredulous. "When? I don't believe you. When was your last date?"

He looks back at the road. "Not quite two years ago," he says. I bark out a laugh. He grins: "But it was a very *good* date!"

We howl together for two miles.

"You should have gone out and had more, Tax," I say after we wind down. "I wouldn't have cared. I mean, I would have been nice to them."

"Maybe," he says, with a wry smile. I start to protest. He says, "Don't worry about it. I probably haven't given up any more of my love life for you than you have for your cello. You haven't exactly kept the parlor full of boyfriends."

"I don't like the idea of having a 'boyfriend.' I bet I never do. It's too much like having a dog." Taxi bought me a puppy when I was about six. I never fed it, never took it out; I only liked it when it took care of itself. He gave it away to the daughter of one of the pressmen at his printer's shop.

Taxi says, "You know, for a while I had the idea you might fall for that kid in your improvisational group, the one who wears the snazzy pleated pants and thin ties, the one we gave a ride to once and dropped on a corner in Georgetown so he could bop into the night with his trumpet case. I remember he snapped his fingers goodbye. And he had a chipped tooth. When I was in high school girls always thought the guy who had a chipped tooth was sexy."

"If you thought that, you don't know me at *all*."

He looks hurt, and says, "Sometimes I don't have a lot to go on."

I look for a new topic quick, and go back to the original one. "So, all the reading and the talking and the cards—did it do me any good?"

He looks over at me. "You can probably answer that better than I can." I start to argue but he goes on: "No, really. You're the one who knows if you're in good shape. If you can be satisfied with most of the things in your life, and can change the things you *don't* like—well, then everything that's happened up to now did you some good."

"Everything?"

"Everything."

I think about this for a minute. Then I ask him: "What about you? Are *you* in good shape?"

He thinks, and smiles a thin one at the road. "Yes. For now."

"Then everything that's happened to you—for example, getting dumped by your wife and winding up with a baby—has done you good?"

He looks at me. "Do you want it straight? No crapola?"

"No crapola." But I shiver suddenly.

He checks the road, then looks at me with simple honesty. "Having you is the best thing that ever happened to me."

I look away. I watch the soybeans; my salty eyes sting. I can't take this. I don't want to know. I don't like being wanted like this, now.

16

So Dzyga had found *me*.

There could be no doubt he was the new cello teacher at the Phrygian Institute—the mention of the Szymanowski wrapped it up. I had never performed that piece except in the conservatorio, and I had never recorded it, except that once, and the only tape ever made had gone over to the studio in Poland where Dzyga had recorded eleven months before.

He could have come over pretty quietly—one musician with spotty credentials wouldn't be noticed among the undesirables getting out of Poland. Probably Dzyga had been teaching for many years in the Soviet Union; perhaps he had also been composing, which was why he fit in with the Institute's new approach. Maybe he wasn't allowed to compose for performance in the Soviet Union. Maybe that's one reason he decided to get out. So he went to Poland, which seemed pretty out of the way at the time, and he risked his cover by making a record he could use to get a job in the West, and he got out of Europe and found this new school, where maybe somebody knew who he was or maybe not—but he got a job.

I thought they probably knew who he was. People like him don't have to play it low for long in the music world. People like him are put at the top pretty fast, wherever they show up.

Maybe he even had something to do with the school being founded. And something to do with the school wanting to hook the only cellist ever to follow his own high path, which nobody had taken a step on for thirty-six years.

Dzyga wanted me. I couldn't help thinking that: he wanted me there.

I couldn't even hold myself at that point—I went on to think even weirder things. I had the idea that maybe my tape had had something to do with his decision to come over to the West. He had been wavering, trying to decide, and then out of America came this astounding tape. Were there really such students in the West?

I wasn't even humble enough to stop there. My idea, deep and proud, was that he didn't want to come over to surround himself with half a dozen apostles. He wanted to come over to walk that high path with the *one*. He wanted *me*. Our stories shared too many lights and shadows. It was eerie. It was an end to being alone. It was destiny.

Of course, in these crazy thoughts, I decided that from nobody else in the world could I really learn what I deserved. Juilliard seemed like an old city high school. What could I get from Juilliard?

I wanted to do the Phrygian audition, and I didn't want to tell Taxi about it. But it would be tough to fake a concert date in California; I had to think of some excuse for getting out there.

Then the idea of linking the audition with the long-promised visit to my mother popped into my head one night. It was perfect.

For Juilliard, Taxi had taken care of the living arrangements very carefully. He found me an apartment in what used to be a carriage house behind the large brownstone of a Columbia professor who had been one of his teachers in college. He planned on coming up every Thursday and returning every Monday, which would mean he spent half the week with me, and the rest of the time I would be pretty much on my own but under the wing of the prof in the house. He had a housekeeper who would make my meals and all that. But in California, what would I do? Taxi would never let me be on my own that far away, and he couldn't skip across the country for three days a week.

Ah, but if I can live with a Columbia professor, I

can live with my mother. She'll pretty much have to put me up, the poor old pothead squaw. It might mean she has to clean up her act a little; it might be inconvenient as hell. But it will teach her something she needs to learn: you just don't throw Sibilance T. Spooner away.

17

Last night we drove far and camped late. I was beat. No sleep the night before, up at 6:12, through that melodrama in Des Moines, long hours practicing, and then another few hours of bus-bouncing. Beat.

But after setting up the camp and feeding me a fast meal, Taxi whipped out the Martin and sat on the edge of the bus, as if he had been begged by a bobbing-headed kindergarten class to sing songs. I couldn't believe it. But I was too tired to put up a fuss, at least for the first few minutes. After that I fell asleep, and I suppose he stuck me in my sleeping bag.

Then this morning, after breakfast, he got out the guitar again. I groaned, but decided to stand it, and I made it through a few songs. But then he just went too far, breaking into a song that wrapped all of this groovy good-time philosophy he's been talking about into a perfect package. I heard it through once: four pompously hung chords, which, actually, were de-

rived from the famous sequence in the *1812 Overture*.
The lyrics were the worst yet; they burned me up,
especially one phrase:

I'm gonna take you, girl, and hold you,
And do all the things I told you in The Midnight Hour.

Taxi was really wailing; I could see he was going
to play it through again when he went back to that
four-chord opening. I put my pinkies in my mouth
and whistled, the way Jake Pettigrew taught me in
fourth grade when we both had to stay inside for a
recess. Taxi looked up, smiled and nodded, as if to
say, *Glad you like it—it's a real hot number, isn't it?* and
kept right on singing. I stood up and waved my arms
and shouted "WHOA! Time out!"

He looked up, confused. "What's the matter?"

"Aquarian music," I said. "I've tried to develop
respect in the season of the witch, I've picked up
gilded splinters under the boardwalk, I've been groovin'
with a black cat bone. But it just doesn't light my
fire, Tax. I mean, come on—listen to what you're
singing! Those simple-minded rhymes? That macho
brainlessness?"

"There's a certain unashamed *innocence*," he said,
"a *directness* . . ."

"Give me a break, Tax."

" . . . And the concept of midnight being a time
for *beginnings* . . ."

171

"I thought we were strictly into the Big Morning here. Hang up the theory, Taxi. And the guitar. From now on, just drive my car, okay? Beep beep yeah." I got into the bus and got on my stool and played scales until we left.

Now we've gone for hundreds of miles without talking. I've been sleeping on the duffels and playing the Bianchi on my stool, hanging out in the back of the bus.

After practicing my Szymanowski for almost three hours, I can't put any more heat on the left hand for a while, so I climb back up front. We're passing through all these weird knobby hills and I just watch. I know Taxi could tell me all about them—what caused them back in the Ice Age or whenever, what layers of soil they have now, what kind of animals live around them for what reasons. Taxi knows all that stuff. I realize I'm not picking up much about the country as we drive across it.

The silence is getting a little thick with both of us up front. I try to think of something to say, but everything I come up with turns into some kind of a cut before I get it out. Finally, Taxi asks: "What was the piece you've been practicing?"

"The cadenza from the second Szymanowski violin concerto."

"That's just a cadenza? All of that?"

"It's a big sucker."

172

"Is he Polish?"

"Who?"

"Szymanowski."

"Naw. Irish. It's an old Gaelic name." Careful, Sib. I take a deep breath and try again to be nice. "Did you like it? Think I was playing it okay those last two times?"

He usually loves whatever I play, but this time he surprises me: "I think you're nervous about it. It sounds like a show-off piece, the kind you always say you despise people for playing."

"You're crazy! It's a beautiful piece of music!"

"I'm sure it is—in its concerto. But like this it's an encore piece."

"How do you know? You never stick around for my encores!"

"Right! Because you *asked* me never to stick around for them, because you feel they're pointlessly flashy and degrading to the program! And that's how this strikes me. Sorry, but it does."

"I don't believe you. This piece is great. You just don't like it because I gave you crap about that nonsense you tried to play me this morning. Admit it."

He's angry again. "I don't play games like that and you know it. Why is it that you can criticize music I make, but I can't criticize yours without having hidden motives?"

"This is different."

173

"Why?"

"You're just *playing*. I've got to *master* this piece, *now*."

"Why?" he says quickly. I've said too much. So I start humming and look out the window and let his question fade, knowing he won't ask it again.

But he does. "Why do you have to master that piece, right now, in the middle of Colorado?" I look at him. He's got a hint of a dark smile on his face, like he's toying with me, and it gives me a shiver. But I'm absolutely sure he couldn't know. I've kept the Phrygian letter in my pocket since the day it arrived. My D file is locked in a drawer in my room. The only thing he could have found is the Polish record, and that would tell him nothing. So I bluff it out with a sudden inspiration.

"Because," I say heavily, "I want to play it for my mother."

In silence we continue through the hills, which turn all of a sudden into mountains. They don't seem as fabulous to me as the Alps: These look dirty and harsh, dramatic only because they are blocking the way. The bus is handling the ups and downs pretty well, though the strain sometimes makes the engine sound like a kid whining for a milk-and-cookie break.

Late in the afternoon Taxi takes out a couple of maps and starts studying every dinky trail-road that we pass.

"What are you doing?" I ask suspiciously.

"Looking for our turnoff." He slows down again to peer at another wrong turn.

"What turnoff? This is obviously the only road that takes us over the mountains."

"We're going to make a stop."

"What stop? We've got gas. We've got food."

He slows down and peers at another road; a pickup truck behind us blasts its horn and swerves around the bus on the right shoulder. Taxi ignores it and nods as we drive on. "We're getting close."

"To what? What could possibly be out here for us? We should keep going. We're going to California, remember? It's thataway."

"I want to stop to see someone. Someone I want you to meet."

"Another weirdo from the Golden Days? Forget it, Taxi."

He frowns. "There are some sides you haven't seen yet. Raoul was a leading political radical and antiwar activist, and he represents some important . . . "

"Important to whom? Give up, Taxi. This archaeology bit isn't working. Isn't it obvious? How many lessons do you need before you pass?"

"The lessons are for you."

"That's where you're wrong."

"Here's our turn. We'll see who's about to learn something." He clicks on his blinker and puts both hands on one side of the steering wheel.

As soon as he starts to turn, the bus cuts off and coasts to a quick stop.

Taxi turns the key. Nothing. "Damn," he says. "It *couldn't* be the solenoid."

I can't hold back any longer; and I break into crazy, helpless laughter. Taxi sits there watching me blankly, hands on the steering wheel. After a minute he gets out and starts fooling around with things on the bus.

I roll down my window and lean out. He's fiddling in the back. "Taxi."

"What?" he says impatiently.

"You won't find the trouble."

He grunts knowingly. I hear a metallic rasp and a muttered "Damn!" as he pulls something out of its place. After some more tinkering he sighs and walks around my side of the bus, wiping his hands. "Unlock the side door, please," he says. "I need the toolbox."

"No you don't," I say. He reaches through my window to get at the side-door lock, but I hold his wrist. He looks at me with a small burn of anger.

"Let go. I need the tools."

"You need a lesson," I say, "and the bus is teaching you one."

He stares, getting angrier. "What are you talking about?"

"It's the soul, not the solenoid," I say with a laugh. "It's the bus's spirit. Telling you that we shouldn't

176

go chase Raoul. It's saying, *No more sad Sixties crapola*. Face it, Tax—this thing, perfect artifact of the period that it is, has come to its senses before you have."

He pulls his arm back and wipes his hands some more, flicking critical glances at parts of the bus and back at me. "You're a funny one to talk about the bus's soul," he says.

I shrug. "I can't argue with evidence."

"Thin air in the carburetor is evidence of nothing but the fact that we're driving at 7,000 feet."

"You're being intentionally dense. It worries me, Taxi. You've never been this way before, at least around me. You've always faced the facts. But for the last few days you've been fooling yourself."

"Oh yes? Maybe you'll tell me how."

"Sure. For one thing, your little plan to teach a crash course in the 'era' that shaped my mother, so I wouldn't hate her for giving me away. So I'd understand her motives. So I'd like her."

"What's wrong with that? If you'd been more patient and interested in understanding . . ."

"It isn't my mother you're trying to make me understand. It's you."

He gapes, then sputters, seriously wronged. I hold up a hand. "I'm not saying you did it on purpose. You may have started out thinking it was to teach me about the old macramé queen. But I figure your nostalgia and your fear took over. Why did I only hear *your* favorite songs, and none of my mom's? Why did

177

I meet *your* friends?" He just watches me. I go on: "I'll tell you why: because *you're* the Sixties person, Tax. You miss it and you want it to be alive. And this trip isn't just toward my future—it's toward your past. I could see that from the beginning. You could too."

He thinks for a second. "Okay, that's my 'nostalgia,'" he says tersely. "You also mentioned 'fear,' I think?"

I take a deep breath and nod. The air *is* thin. And cold, and quiet. I wish some cars would come by and break this up a little.

He watches me carefully. "So—what am I afraid of?"

I look him in the eye and feel my throat closing. "That I'm going to leave you."

He takes it in, and nods. Then, in a voice as constricted as mine, he asks: "And are you?"

It's as if I've forgotten how to make the right sounds for the word—the air won't go where I need it. But I manage to hiss it out for him: *"Yes."*

He nods. The late-afternoon light is making him look clear and fine-tuned, and I have the feeling that I am seeing the definitive Taxi for this instant: the thick sprawl of hair blowing slightly, the grooved face, the grace in his stance beneath the rumpled khakis and baggy SIERRA CLUB sweatshirt. It occurs to me that maybe I am supposed to memorize him this way. I try to memorize him. It doesn't work. He's there, but

178

I can't take him in and file him away. I get scared I'm losing something, and then he moves. He walks out of the light and around the shadowy bus, and gets in.

"Let's go to California," he says, as if to the bus. Then he turns the key.

It starts.

Home

18

Nobody's awake in the house, so we're just watching from the bus, parked up the street. We've been watching for an hour.

It's a pretty nice house, a big brick Victorian place built on a high corner plot at the lip of a hill that dips down toward the city, with a view of the bay off in the distance and a breeze that smells like an ocean, not a harbor. The house has new, widened, smoked-glass windows with no struts, and a genuine slate roof that looks pretty fresh. The yard, which slopes down on all sides to a stone wall that holds it above the sidewalk, has no more bumps than a Haydn string quartet.

Before we found the house, we stopped once on the edge of the city, at a high place overlooking the bay, where Taxi pulled up and got out and came around and took out the Martin. "You needn't listen," he said quietly, as I woke up. "This is strictly for me." Then he sat on the edge of the bluff we were on and sang a pretty nice tune, a little heavy-handed, but okay:

> *I left my home in Georgia,*
> *Headed for the Frisco Bay.*
> *I have nothing to live for,*
> *Look like nothing gonna come my way.*
> *So, I'm just gonna sit on the dock of the bay*
> *Watchin' the tide roll away,*
> *Sittin' on the dock of the bay,*
> *Wastin' time.*

His voice was sweet, and the morning air was cool. A sea gull suddenly lifted up from below the bluff and rose right in front of Taxi and away to the east, but it didn't cry out. After the last verse of the song, Taxi whistled a little improvised melody off the changes, finishing with a chorus that trailed over the water to the city. Then he brought the guitar back and started the bus.

"Nice," I said.

He nodded. "Thanks."

"I mean it," I said.

He smiled. "Of course you do. I haven't started doubting everything you say just because you've had a couple of secrets, Sib."

"I guess I'm jumpy. Meeting my mom. Having that audition."

"I'm jumpy too."

"The song helped, though. Really. Did it help you?"

He thought for a moment, almost frowning. "It's a very *sad* song," he said, and nothing more. We headed down toward an early-open gas station to find our way to my mother.

We were both pretty surprised when the neighborhoods got better, not worse, as we got closer. When we finally arrived at the address, we had to drive past twice to make sure it was the place. Then we parked, and stared.

Finally, I said, "Maybe she lives over the garage."

He nodded, still watching the house.

"Maybe she works there, like a live-in maid," I said. And of course I'm wondering how a live-in maid is going to accommodate a live-in daughter.

We sit and stare at the house. Suddenly in the back a light goes on. "We can wait a while longer, until a few more go on. Maybe she isn't the first one up. Maybe that's the housekeeper or something, and the maid sleeps later."

"Sure," he says. "We can wait as long as you like. We're here, though."

"And so is she, I guess. That much is pretty certain."

"I suppose so."

We don't talk for a while. A little window lights up upstairs: a bathroom. "Taxi."

"Yes?"

"Have you been a great father?"

He laughs briefly. "That's more for you to answer than me."

"No, it isn't, really."

He thinks. "Well, the only way a father can know that is if the child knows it."

"But how would I know it? What are the things a child should be that show she's had a great parent?"

"Happy."

"Be more specific."

He smiles slightly. "Happiness is very specific." He looks at me. "Parents are pretty slippery, when it comes to judging how good they are."

"It seems like kids are, too."

"That's right."

I think for a minute and look at him. "Make a deal? I won't judge you if you don't judge me."

He laughs. "No deal."

"Why not?"

"Because I *want* to judge you. I want to feel you're a great kid. And I want you to judge me. I want you to look back and decide I was a great father." He shakes his head, smiling at himself. "That's not why I've done what I've done, but I'd be lying if I said it wasn't important. Especially now."

He nods toward the house. Two more lights are on, one upstairs, one downstairs. The sky is going from bluish gray to yellow-blue, too. "You can go anytime, I think."

186

"*I* can go? Aren't you coming?"

He looks at me with a slightly puzzled smile. "Why should I?"

"To . . . to . . . to introduce me." The idea of going up there alone never occurred to me. "To explain our coming. To . . . "

"You can introduce yourself. And you can explain our coming. You know why we came better than I do—it was your idea, remember?"

"Yes, but you always *offered*. You had a reason why you thought I should *want* to come someday, didn't you?"

He shakes his head. "No. I just wanted you to know the road was open, if you discovered reasons of your own. And you did. Even if you can't pin them down at the moment, you found some. Even if you think it was just for an audition."

A paperboy rides by on a ten-speed bicycle and flips a plastic-wrapped newspaper in cartwheels through the air. It lands smack against the screen door on the front steps. As I watch, the door opens just a crack, and a hand shoots out and snatches the paper. That hand makes me shiver.

"All right," Taxi says. "I'll come with you."

We walk down the sidewalk and up the steps. I glance into the front room, lit behind the modern windows, and see a corner of bookshelf with a big speaker on top of it. *A good omen*, I think, grasping at any sign—*there's music in the place.*

We walk onto the small porch; Taxi hesitates at the door; I push ahead of him and knock six times quickly and loudly, the same rhythm and volume as my heart-beat.

In a few seconds the door opens and a dark young man stands there in pale-yellow pants and a double-breasted blue blazer, unbuttoned. He has on a pale-blue silk shirt open at the collar and a loosely knotted burgundy knit tie, very skinny. His hair, very dark, is wet.

"Good morning," he says politely. Taxi and I say nothing for an instant, and the man goes on, as if to help us out. "You're either lost or you're Mormons. In which case, you probably think *we're* lost." He smiles. If I were a Mormon I wouldn't be offended. But I wouldn't feel exactly invited inside, either.

Taxi speaks up. His voice starts out too quiet, and then when he realizes this, he makes it too loud. "We're not lost," he says. "At least, not if we can visit Connie here."

"Connie," says the young fellow, looking at us. "This is interesting, isn't it? Come in." He steps back and opens the door. We walk into a hallway yellow with old oak and maroon with an Oriental rug. Through an arched doorway off to the left, I hear Shostakovich's eighth quartet playing at very low volume. An interesting choice for a morning; but it may be a radio.

"Who shall I say?" the young man asks, again with

perfect politeness. Taxi says, "Please tell her Cabot and another visitor."

The young man's eyebrows arch slightly. "My," he says. He looks at me and bows slightly. "Perhaps it would be best if I let you introduce yourselves?"

He seems to want an opinion, so I nod. He motions us with his hand to follow him, and walks down the hall and off to the right.

We enter a small room with a bay window and a lot of steel-framed photographs of architecturally snazzy buildings on the walls. There's no furniture except a small rolltop desk in a far corner and a long antique wooden table running the length of the room under the window. There are six ladder-back chairs. In one of them sits my mother, over a pale aqua teacup. She is reading an advertising section of the newspaper. A set of small speakers hung in the inside corners of the room ooze the Shostakovich, which must have covered our steps. She doesn't look up, but there she is.

The young man coughs faintly. My mother looks up. Her glance whips over me, to Taxi, back to me, and finally rests on him. "Hello, Cabot," she says, without a hitch of drama or surprise.

"Hello, Connie."

She nods, putting the paper down quietly. Her eyes are unbelievable—very bright blue, but so light in color that their brightness seems impossible. She's small—I notice that her wrists are thinner even than mine, and her hands are elegantly tiny—but her face seems slightly larger than it should be. It seems slightly prettier than it should be, too—her nose is finer, her cheekbones clearer and rounder, and her mouth more expressive than you'd expect to find in a face without much humor. I bet she never really laughs.

She smiles slightly at me. It's an intimate, wry smile, more intimate and wry and exactly right for me than it could possibly be. She must have just smirked and gotten lucky.

"I'm sorry to be so long in greeting you," she says. "It's just that I'm trying not to begin by saying *Well, well*."

"Maybe I should say it. Well, well."

She raises her eyebrows and looks back to Taxi, shifting to a different cool smile. "What brings you?" she says. She's good: in her tone is a fine edge that lets Taxi know this is an outrageous impropriety on his part, while preserving me from feeling the same.

Taxi says, "Every year I've let Sib know that I would take her to see you, once, whenever she asked." He nods at me. "She asked."

I don't acknowledge his nod; I'm not sure what to say if my mother wants to know why I asked. But she doesn't; she's not through with Taxi yet. "You might have dropped me a line," she says.

"I might have," he says.

They stare at each other. Finally she says, "Oh, yes, of course. But, no. You're wrong," and shakes her head. He says nothing.

"Your father was afraid that if he let me know you were coming, I might disappear," she says to me. "I'm sure he didn't tell you anything like that, though?" It's a question.

"No."

She looks back at Taxi. "I suppose you'd better have some breakfast." Again, her manner strikes a clear distinction: Taxi needs to be invited, but I don't. I'm torn between feeling resentful of the dividing that's going on and relieved. I'm tired—suddenly I feel like I've been tired for sixteen years—and I want to just accept this small delicate welcome that's being offered.

I can see Taxi can't decide whether or not to fight it. If he looks at me and says, "Shall we stay, Sib?" or something like that, it will force the issue. I honestly don't know what I'd do. But he decides to let go. "No thanks," he says. "I'd better be going." And suddenly the link is broken. We're not a twosome anymore, Taxi and I.

"I'll bring your things from the bus," he says to me.

"Martin will help you," my mother says.

"No," I say, "wait a minute." I turn and look at Taxi. He's barely holding himself together, like he's about to start shaking all over. His face is so serious it frightens me. I start to panic that this moment is going to pass by without my feeling it, that I'm standing here cool and adjusted, while my father goes away.

191

This is your father, I tell myself, *this is important.* But the word *father* doesn't mean any more to me than *mother,* really. All I see standing there at the door is good old Tax, and I know too much about him for a title to tell.

"I'll come with you," I say. I look back at my mother. "I'll bring my stuff in myself."

"Fine," she says. She and Taxi don't say good-bye.

I can't think of what to say as we walk in step up the street to the bus. I start to cry. Taxi takes my hand, and holds it like a boyfriend while we walk.

When we get to the bus, he says gently, "I'd like to come to the audition, if that's okay."

I nod, snuffling.

"So this isn't really goodbye," he says, touching my cheek. I nod.

He opens the side door and gets out my duffel and the Bianchi. "Taxi," I say. He puts them down and looks at me. I sniff, and let out a deep breath. "I'm sorry I messed up the trip. I should have just let it be, you know, our big trip together, without being so picky."

He shakes his head. "It's me who should have let it be our trip," he says. He glances at my stuff and the house. "But we got here anyway."

"Yeah. I guess I'd better go in." I sigh, and pick up the Bianchi and duffel. I look at him. "I'm glad you want to come to the audition."

He smiles wryly. "I'm glad too."

"Did I mention the name of the school?"

"The Phrygian Institute," Taxi says. "I'll find it."

"Okay. See you there."

"Right. 'Bye." He gives me a last look, then climbs into the bus. I watch him drive away, and then I walk back to the house.

The young guy opens the door and takes my things, raising his eyebrows at the Bianchi. "I'll be careful," he says, before I can phrase a warning. He takes the stuff upstairs, and I walk back into the dining room. My mom puts down the paper again.

"Was it a hard trip?" she asks.

"No."

"You drove all the way in that old bus that just went by?"

"Yes. It's a great bus."

She nods, frowning slightly. At that moment the young man comes back into the room, carrying a large silver tray with covered dishes on it. He smiles at me and puts the tray down on the far end of the table. Then he stands and raises the covers one by one: "A cream cheese omelette with artichoke hearts, neatly divisible into thirds. Smoked salmon, caught by your hostess and tinned at a fishing camp in Oregon, by the way, which she may neglect to tell you. Sourdough bread." He looks at me. "Perhaps you've never tasted sourdough?" he says, with the slightest condescension.

"Not by that name. But there's a yellow bread I've eaten in Yugoslavia that I think is probably pretty similar."

"Oh!" he says, "Touché," and leaves. My mother laughs and flicks me an approving look.

"When were you in Yugoslavia?" she asks.

"I was there twice," I say. "The past two summers."

"I see," she says. "Did you like it?"

"It's a beautiful country," I say. "Great bread."

She smiles a short one. The young man comes back in with a steamy silver coffeepot, holding the handle in a white-and-blue-striped linen towel. He puts it down and motions for me to start. I say, "Thank you. . . ."

My mother says, "Oh, excuse me. This is my secretary, Martin Meriweather."

He smiles and bows. My mother says, "I believe I heard your father call you 'Sib?'"

"For Sibilance," I say, blushing. "I changed my name when I was eight. It's Sibilance T. Spooner."

"There's no need to blush. Sibilance is a lovely name. I'm relieved no end that you haven't been cursed all these years with the name I gave you. You thought of Sibilance yourself, I gather."

"Yes."

She nods. We eat. Martin and I do most of the talking. He asks harmless questions that draw out good information: how we traveled, how long it took us, where we stayed. My mother looks up twice with her eyebrows hiked—once when I reveal that we camped every night, and again when I mention that we drove straight from Colorado with only a couple of nap stops.

After the food is gone and we're all sipping coffee, my mother says, "You play the cello."

"Yes."

"You must play it seriously; any young woman who would cross the country with a musical instrument larger than her clothing bag must have musical priorities."

I smile and nod. I'm a little nervous: I haven't figured out how to bring up the audition. I'm not feeling as blunt and brazen as I thought, and suddenly I don't want it to seem like the only reason I'm here is for music school.

I leave it to my mother, and she gets to the matter with half a dozen very good questions about my music. I tell her, blushing, that I was invited for a scholarship audition at a new music school in San Francisco; its programs interested me, though I have been planning to attend Juilliard this fall. The audition is on Saturday.

"May I ask the name of the school?"

"The Phrygian Institute." She smiles. I ask: "Do you know it?"

"I'm on the board of directors." I gawk, and she hastens to qualify: "Purely in a business capacity—

the school was a client of mine in real estate and architectural matters. The chairman asked me to lend my experience in certain narrow areas. I'm not a musician—that *would* be too frightening a coincidence."

She pauses, puts down her cup, and moves a little forward in her chair.

"I know of course that the Institute is forbiddingly high-minded in its recruitment of faculty. And students." She fixes me with her pale, luminous eyes. "In other words, you must be dynamite." She smiles slightly. "True?"

I have to smile. "True."

"Were you in Yugoslavia for music?"

"Not in Yugoslavia. That was kind of a side trip. I was in Europe for music, though—a couple of competitions."

She nods. "I've heard of those. Some of them are very prestigious and, if I'm not mistaken, hotly contested. Also very well judged." I say nothing. "Did you win?"

"Yes."

"Both?"

"Yes. Plus one the year before."

She settles back, looking at me. "So you *are* a star," she says, breaking into a deep smile that is proud but not possessive—not even maybe as much as I'd suddenly like it to be.

There are tears in my eyes. Did I ever really think I would answer something like this by saying, *And what are you so proud of—you kicked me out*? I can only say: "Maybe I should have kept the 'Starness' part."

But she's past the gushy moment. "Of course you shouldn't have. It's a ludicrous name. I was a fool then. The name I stuck you with is only a sign of many mistakes. Do you understand that?"

I nod.

"Good." She runs a hand through her hair. It's pretty hair—brown, thick, wavy, very spontaneous looking but stylish.

She notices my scrutiny. "Shall we get you a haircut?"

"Do I need one?"

"More, probably, than you need new clothes, but we'll get you some of those, too. Why did you bring all of your things in a duffel bag? They're certain to be ruined. Don't you have any luggage?"

I hem and haw; she nods and cuts me off. "Of course you don't," she says. "What would your father know about luggage? I'm sure he thinks duffels are *fun*." She gives the last word a Taxi-like inflection, and I smile for a flicker; she sees the smile and returns it quickly.

"Actually," I say, "he does know about luggage. He bought me some great luggage. I just . . . didn't bring it." I yawn.

"You're exhausted," she says. "Let me show you your room."

She takes me upstairs and down a hall to a nice,

cool, dark room with a bath. The cello case is standing in one corner; people who don't play cellos always think they have to be upright even in their cases. The bed is turned down and looks terrific. My mother tells me that my clothes have been unpacked and are now being "salvaged" by Harriet, who must be the maid. After my nap we'll go shopping in town.

I thank her again, and we stand staring at each other. She is looking hard, trying to figure something out. Finally she says: "I can't figure out where you get it. Geniuses get *some* spark from *some*where. But I haven't a musical bone in my body, and neither does your father."

"You have Shostakovich quartets," I say.

She gestures impatiently, still studying me. "Martin buys my records. I have no idea."

"And my father's a little musical." Her eyes get critical. I add, "He plays the guitar."

She laughs. "Oh, certainly. Does he still? All those soul classics and hippie political anthems. That *must* be it—I'm certain you derive your world-class brilliance from the genes that boogaloo to 'Shake a Tailfeather.' " She bends forward and quickly kisses me on the mouth. "Have a nice sleep."

19

I sleep for a long time, dreaming of sleeping in the bus through night and day, not being able to wake up, hearing Taxi crying and trying to wake up but falling back. And then, after a long time, I try to surface again, and hear Taxi—but this time he's singing, and as I struggle and fall back and surge again toward waking up I realize that he's not singing one of his songs—he's singing, note for note, with a distant sound, my own version of the Prokofiev Cello Concerto, exactly as I recorded it under Rostropovich three years ago. It gets louder and closer. I sit up.

But the music doesn't recede with the dream; it's coming up from below. The record. My mother has one of my records. Once a long time ago I thought of this possibility—my long-lost stranger mom would happen across a record in a store, be mysteriously drawn to it, take it home, play it, find herself seized by a rhapsodic sense of familiarity and mystery, and listen over and over, knowing something but not enough. . . . It was romantic comic-book stuff, part of the dizziness of making your first recording. I've made a dozen now, and it's been a long time since then.

I hop out of bed. My clothes were returned while I slept, washed and ironed. What do you wear to go shopping with your mom? I throw on a pair of cotton drawstring pants and a loose jersey and tennis shoes.

I go downstairs. My amazing third movement is booming. A good sign: people who listen to music loud usually *listen* to it; they don't keep it down so they can chat over it or vacuum the carpet. But when I reach the living room, there's only Martin. No mom. He wiggles his eyebrows at me, not breaking silence over the music, which, even in my huge disappointment at finding him alone, I recognize as a courtesy. His elbows are propped on the chair arms, fingers pressed together in a relaxed arch; in his lap is the cover of the record. It's brand-new, with the shiny shrink-wrap and the price sticker still on. My mom hasn't been rhapsodizing over *this* one for long, lonely years.

The concerto ends. I go and sit in a chair that looks like cloth, but turns out to be fiberglass.

"A beautiful record," Martin says.

"It's a great piece," I say. "Maybe the best cello concerto, though I like Frank Bridge's a lot too."

"Better than the Dvořák? The Saint-Saëns? The Elgar?"

"The Prokofiev and the Bridge are more open to discovery." I shrug. "I like to discover things in a piece."

He smiles and puts the record jacket aside. "That's a watchword I shall take for my efforts as a listener, too."

"What do you do, Martin? I mean, my mother said you were her secretary, but . . ."

"But that can mean anything, right?"

I nod.

"In this case, 'anything' isn't much," he laughs.

"Is her work like real estate?"

"It *is* real estate," he says, "and a few related things. She started a new field out here—architecture brokering. It's complicated, but she has developed the means of hooking up the right architect with the right landowner with the right building client with the right contractors on the right piece of land, for all kinds of construction projects."

"Are you a real estate expert too?"

"God, no." He laughs, getting up from the chair to take the record off. "I'm a glorified dropout from architecture school. Your mother discovered me being clever and critical on the local arty-party circuit, and thought perhaps she could turn me into something useful." He takes the shrink-wrap off with a crackle and sticks my record away.

"Was *she* ever an architect?"

He shakes his head as he switches off the electronic things on the stereo. "No. She's a businessperson." He turns to me and lifts his hands: "Shall we go shopping?"

I stay seated. "I thought I was going shopping with my mom. No offense."

"No offense at all. Your mom intended to take you herself, of course. But an important client called and demanded a couple of hours of her time. It's part of being a businessperson." He smiles warmly. "I much prefer being a dropout. Shall we go?"

20

In a little joint on a dirty wharf we eat a peppery seafood salad, sourdough bread, and freezing-cold white wine served in school-cafeteria-issue tumblers—earning Martin a few points for not taking me to some *nouvelle cuisine* spot. Then we start hitting some stores.

At about the third store I ask again what my mother is doing. I'm feeling grumpy. I don't like to shop.

Martin, chin in hand as he studies to decide whether I should try on a silk print dress in mauve next, or crimson, murmurs that we're meeting her later; she's showing someone a site.

"A site," I say, "or a sight?" I do an awkward little spin with the mauve dress.

"Try the crimson," he says, handing it over. "Mauve is too trendy for you."

Three stores after that I have four new dresses, three pairs of shoes, a little flat handbag I couldn't carry anything but an emery board in, a flashy little rain jacket, some pants and some sweaters and a sweat suit, and a bunch of silk scarves that tie the whole mess together. It's all nice stuff, and Martin seems to know fabrics and cuts and dyes and all that junk better than the people in the stores—all of whom he also seems to know. He picks almost everything out. One of the sweaters is my choice; the rest is his. Several times he says stuff like, "Your mother wanted something like this," and I wonder when she had time to come up with all these instructions.

By this point I'm worn out and bored. At first I kind of liked coming out of dressing rooms in good dresses and having Martin smile and nod as he looked from my shoulders to my knees, but now I just want to sit down. We head for a park, and I collapse on a bench. There's a fountain, and some people playing guitars off under some trees, and a guy in a black suit doing mime. Martin even knows *him*: The guy breaks character in the middle of mocking an old bag man just to wave hello.

"You know a lot of people," I say.

He sits beside me on the bench. "I was out discussing cayenne and prawns with chefs, ramie versus silk with tailors, and wrist control with mimes when I should have been studying heating ducts and electrical wiring plans." He sighs. "Every person I know is evidence that I deserved to be kicked out of architecture school." He looks happily out over the park. I watch him, unable to decide if he's nineteen or twenty-nine.

Then he looks over at me. "Of course, if you're a genius you can withstand temptation."

"If I meet any geniuses I'll ask them."

"Do you really feel that modest?"

"I'm not modest at all, if you mean by modest that I deny my talent. I'm a great cello player. But that doesn't mean I like weird labels people use to set me off from everybody else in the world."

He nods, and leans back. "Of course not. Sorry. I'm talking like a smoothie dilettante."

After a minute I ask: "Are we meeting my mom here?"

He looks at me, amused but sincere. "Are you having such a bad time with me?"

I blush. "Oh no, not at all, that's not what I mean. . . ."

He's laughing, and he leans over suddenly and gives me a kiss. A very light kiss, his soft mouth neatly to mine, no big deal. But my chest, suddenly, is full of a bird flying in the dark. I sit kind of dazed. Martin sits back and resumes talking. I mumble things now and then, but mostly I stare at his lips and eyelashes in the low sunlight. I feel calm, almost *relieved* by this: I can sit here and stare at this guy's lips, and not feel rude or silly.

At least until my mother arrives.

She walks up all of a sudden, gives me a kiss as I

look up, and lays a hand briefly on Martin's shoulder. The next thing I know, the two of them are looking over my purchases. My mother holds her hand nicely on my back and nods curtly at several items; says, "Oh, good, very good idea" at one of the scarves; frowns at two of the sweaters until Martin says I picked out one of them; and claps when he pulls out the crimson dress.

"That will look wonderful for your audition," she says.

"Whoa. I'm not going to wear that for my audition!"

She looks surprised. "Why not?"

"I wear black when I play."

She smiles. "Not in California."

"If I happen to be playing in California, I wear black in California."

"But Sibilance, black will look ridiculous at the Phrygian Institute on a Saturday afternoon. Believe me, California is different from New York or Budapest."

"*I'm* the one who has to be comfortable, and I'm comfortable in black."

She starts to say something else, but Martin says, "I'm sure you look great in black." My mother stops, looks at him, and drops it.

Then we're up and walking and my mother is asking me how I liked the stores and the parts of town they were in, and I'm telling her I liked them fine,

206

nice chitchat, nice walk, nice evening light and birds heading into the trees. I realize I never talk like this—I never cruise along anywhere. It's fun right now, with Martin on one side of me and my mother matching me step for step with pretty strides on the other, in San Francisco where if you wanted to you could play Bach in red.

After fifteen minutes, we reach the BMW Martin and I came in. My mother and I are going on to her car, and then out to dinner alone. I try to hide my unexpected disappointment at this as Martin opens his door and tells us goodbye. But my mother holds him up by asking where we went for lunch. Martin tells her the name of the place.

She looks at him with a pop of sternness. "Oh, why on earth did you go there? I thought—why didn't you go ahead to Mercutio's? Oh, really, Martin."

Martin smiles and shrugs, and holds her eye. "I thought Sib would enjoy it more. I thought it was . . . more genuine."

My mother looks at me apologetically. "I wanted him to take you to a marvelous place, one of the city's . . ."

"I liked this place better," I say. "This place was perfect."

She looks quickly back and forth between us, then shrugs. "All right, fine. I'm glad you had a nice lunch. All the same, we'd better go." Martin and I take a quick look at each other; he winks.

21

My mother decides to take me to Mercutio's for dinner, which means we have to go home and change clothes first. I wear one of my new dresses, a brown one I didn't care for when we got it; now I notice it does look pretty nice. I put on some new shoes, too.

When I show up downstairs, my mother looks at me with her eyebrows raised in approval. "Oh, honey— you're *lovely*!" I blush and look at a plant and pull a leaf off. She says, "That's such a good color for you."

I look down at the dress. "Martin said it was exactly between my hair and my eyes. He said it showed that 'Brown is beautiful.' "

"Did he?"

"Yes." We look at each other. She's wearing a satin dress of lavender that picks up the color of her eyes. "You're beautiful too, Mother."

"Well," she says. "Shall we go be beautiful where some other people can see us?"

The drive is terrific. My mother tools her Mercedes sports car with a lot of zest, but smoothly. We spin along the top of one hill with the harbor on one side and intricate toy-looking streets on the other. "It's a thrilling city," she says.

"Have you lived here ever since . . ."

"Yes." She looks at me as she shifts into a higher gear on a straightaway. "How were you going to put it just then?"

I shrug. "Ever since my father and I left."

She nods, and shifts again. "Did he tell you why you left?"

I look down a street. Somewhere near the end there's a banner across it, and extra lights, and people are walking or dancing. "He said the two of you decided it would be best that way."

"Is that what he told you? Really?" She seems truly surprised. A minute later she asks: "Is that all he said, or did he give you that as a general description, with more specifics, to fill in?" She brakes for a yellow light.

I look over at her. "Why don't *you* give me more specifics, to fill in?" She returns my look. The light turns green.

"The only specific was that I wasn't ready to be a mother," she says.

"Ready in what way?"

"Ready to . . . give what I had to give. What it looked like I would have to give."

209

"And you thought my father was ready to give it?"

She flashes me a quick annoyed look; she wishes I'd keep the two of them apart, I know. "Not . . . well, he seemed at least better placed to try."

"That's not an answer. You can't have it both ways, you know. You either thought he was going to be a better parent, because he was more ready—or you thought he was convenient, and dumped me."

She smiles slyly. "You're going to burn me a little, aren't you?"

"I'm only bothering because I like you," I say. "Frankly, I didn't expect to bother—to try to understand, to hear you. Now I've got a lot of things to work out very fast. I have to get people to tell me the truth. If you're going to try to make things look cooler your way, I'll turn up the heat."

"I suppose your father didn't try to make things look cool his way?"

"No. And I bet you're not really surprised at that, are you?"

She starts to say something serious and defensive, but stops. Then she sighs, and laughs. "All right. No, I'm not surprised your father didn't paint a sweet picture for himself. One of his more . . . troublesome traits was his lack of either the ingenuity or the ego that drives the rest of us to construct our own versions of the world." She shakes her head. "He was very hard to engage in battle." She looks at me. "Battles

210

aren't always bad. And I bet *you* aren't really surprised at *that*. Are you."

"No. I fight a lot."

"Not, I bet, with your father."

"Right."

She waits, then asks: "Whom do you fight with, mostly?"

"The people I want to be my friends."

She looks at me and smiles a pretty good one. "Then I hope we have some great fights."

I smile back. "So which was it?"

"Hmm?" She looks over breezily, feeling a little too smug about being fightin' buddies while Taxi's off the field as a boring noncombatant. "Which was what?"

"You thought my father would be better, or you dumped me on him for convenience."

"Oh, shit." She doesn't look at me. "That's a pretty immature question, and your insistence on it is even more immature."

"What the hell. I'm a kid, you know?"

"You're no kid," she says. "Let's agree about that. You're a woman."

"At least I was a kid sixteen years ago."

She nods, and sighs. "All right. Here it is." We pull into a gravel parking lot whose perimeter is strung with dim Chinese lanterns in white and light blue. She turns off the engine and swivels to face me. "I

211

was afraid. I was afraid I didn't love you. At all. I stayed up all night with you the first night. So did your father. I hated it. He didn't. Hell, he took care of *both* of us, and for that I hated *him*. I hated the thought of ever doing it again, much less doing it every night for a year. I didn't want to give you milk from my body; I wanted my body back for myself, after you had had it from inside for nine months. I wanted to drink brandy again. And coffee. I wanted to sleep until ten, sometimes. I wanted to sleep until ten with a *man* sometimes. When you cried, I didn't give a damn why—I wanted you to shut up. Your father had the investigative spirit. I'll give him this—his response to you was very fine. He read your screaming little mind. Within four or five hours he already knew what three different kinds of crying meant—what we should do to make you comfortable according to the way you 'expressed yourself.' " She blows out a raggedy breath. "More power to him, you know?"

"You decided you didn't have the feelings it took to put up with the crap."

"Exactly." She hesitates for a second, then reaches into the glove compartment and pulls out a pack of cigarettes. Camels. She lights one, exhaling smoke roughly. "Hell, you have no idea what it demands. . . ."

"You don't either, you know."

She looks at me fiercely for a dangerous second, but then she shrugs. "You're right. But at least I sensed it then—believe me, a new mother can feel the years stretching ahead. And I didn't think both of us would do well during those years if I was bitching at every sacrifice. I'm not brave, Sibilance, and I'm not generous. Your father wasn't exactly brave either, but he had no hesitation about taking you on. That first night he said, *I'll feed her, every night; you sleep.* I thought at first he was being grabby and saintly, getting an edge, but then I realized I just didn't care enough to fight for my own part of the matter."

"Of the 'matter'? You mean of me?"

"I guess I do. I let him do it, and he did it so well. That's what really gave me the idea that . . . perhaps I didn't have to put us both through all that mutual misery. You and me, I mean." She takes a puff and blows the smoke out quickly. "I don't think he was a better human being than I was—though I went through a lot of guilt, of course, for a while—I just think it was a matter of constitution, of personality. I *minded,* he didn't."

I suddenly wonder if all the times I've said, *Taxi doesn't mind* or *Taxi's feelings don't get hurt* have sounded as wishful and unlikely as this. This isn't the time to worry about that, though.

"Thanks," I tell my mom. "Thanks for telling me. It sounds scary to me too. I don't know—maybe I'd

213

get out of it like you did." I shrug. She frowns. I say, "Can we go eat?"

She nods, still frowning, and flips her cigarette out the window, still lit. I take her arm for the climb up some redwood steps to the restaurant. After a second she leans against me and squeezes my arm against her ribs. She's skinny, too.

22

We're waiting for dessert and coffee. My mother offered me a liqueur too, but I didn't take it. It's been a nice meal, lots of casual talk, mostly about me and my music, and her and her architecture work. Nothing profound.

I ask her: "Is Martin your lover?"

She looks amused. "It's obvious you don't know San Francisco. Anybody who did would take it for granted that a young man as pretty as Martin, with good clothes and an arty profession, would be gay."

"He's not gay."

She arches an eyebrow. "How do *you* know?"

I shrug and blush. "Is he your lover?"

She laughs. "I'm enjoying you," she says.

"Good. Me too. What about it? He was at your house pretty early in the morning."

She raises an eyebrow. "He sleeps in the study many nights. We work late when we have a big project."

I nod, looking down at my salad fork. "But . . . don't you . . ."

She smiles slightly, amused. "Yes, I *do* find him attractive. I think he's beautiful. That's what you were going to ask, isn't it?" She smiles all the way now, and shakes her head. "He's also a very beautiful architect. Who needs a friend and a boss more than a lover."

As if on cue, the desserts arrive. After a couple of bites she says, "How much about lovers do you know?"

"You put that very delicately," I say.

"It's a delicate matter for two women to talk about."

"Okay," I say. "I've had one lover."

She watches me easily and eats her amaretto torte; I take a slurp of coffee and pick a piece of apple from my pie.

"In Rome last summer?"

I look at her. "How did you know?"

She smiles. "Rome is a romantic place."

I shake my head. "Not in the middle of a musical competition. I never really saw Rome. No pasta dinners in street cafés with candles dripping out of wine bottles. No riding in those little boats with guys in striped jerseys poling you around."

"That's Venice."

"See what I mean?"

She laughs. "I gather then that your fellow was a musician?"

216

"Yes. But we didn't meet in the usual way people do at those things. There's always a lot of nervous sex in the air; it's high-strung and goony, and I stay clear of that whole scene."

"You're very pretty, I hope you know."

"I'm also tall and sarcastic. Boys don't like that. They like round little girls who giggle. Plus I'm a mean competitor. I stay away from the places people go to watch each other. I'm busy. I have work to do."

"Practice. It must take endless practice before you perform."

I eat some pie. "Not really. I'm not a freak for practice right before a recital. I figure I brought the music with me; I'm not going to pick it up in the last six hours. Mostly I just keep my hands in shape. But I do spend a lot of time thinking."

"About the piece you're going to play?"

"Well, actually, I spend a lot of time thinking about some piece that's in another idiom altogether. Sounds weird, but it keeps me fresh for the piece I play when I come to it, because of the contrast. Think about Vivaldi all day and you can't *wait* to play Samuel Barber, even if up to that point you've practiced and analyzed the Barber until you hate it."

"That doesn't sound weird at all," she says. "It's simple and clever. Go on."

I take a breath, and I do go on, further than I've

217

ever gone with anyone else I talk to. It's *easy*. I feel no twinge of hesitation, no thrill at showing off. I'm just talking to another woman. And she happens to be my mom.

"One day I'm walking through the old building they gave us to practice in, and I open the door of a practice room and find myself face to face with a young guy holding a cello. I say *Pardon* in French because that's the universal language most of the time, though I could tell his face wasn't French. I only saw it for a flash, looking up from his music. It was round but kind of tough, with big eyes and long eyelashes and dark heavy eyebrows. A thin mouth. It sounds like some dumb cliché, but he looked a little like a gypsy. I've seen gypsies in Europe, real ones, and I think they're fantastic. I didn't think this kid *was* a gypsy—he just had a little of that look.

"I closed the door and started to walk away. But I had to sit down. My chest was so full of air I couldn't get any more in, except in stabby little breaths. Very weird. It had never happened to me."

She smiles. "Congratulations."

"I had a lousy practice session. I kept getting thirsty. When I knocked off, I went by his practice room and heard a Bach suite coming through the door, with great tone but uneven pitch, and I stood there and imagined his hands making the notes on the neck.

"It drove me crazy for two days. I managed to

cruise through the preliminary rounds of the competition, but when I saw him a couple of times walking around, I felt awful and played terrible cello, for me. Borderline. I could tell the preliminary judges were dismayed. The year before, after I won Brussels, everyone wanted me to fall at Prague—it's common, see: At first no one wants to believe you're as good as they thought you were. But in Rome I had won twice and was the champ, and they *wanted* me to win.

"Finally one day I saw him sitting by himself at a table in the café garden of our hotel. He was drinking a French mint soda. I didn't think about anything—I just bolted up to his table and collapsed into the chair across from him.

"He looked up and smiled.

" 'You do not play as well as they said you would,' he said to me in French. 'Why is this?'

" 'I think it's because I want you to kiss me,' I said. 'I can't figure it out. I never want *anyone* to kiss me.'

" 'This I can see,' he said.

" 'Well?' I said. 'Will you? I mean, sometime—not right now.'

"He stared at me. 'If I kiss you, will you feel better?'

" 'I guess so,' I said. 'God, I *hope* so.'

" 'And then you will play better. Perhaps as well as they say?'

"I shrugged.

"He smiled. 'Then I will not kiss you.'

" 'Well, shit!' I said angrily. 'Why not?'

" 'Because of the boot factory,' he said. 'If I kiss you and make you feel better, you will play very much better than me, and I will go back to my home in Hungary in disgrace and be put to work in the boot factory in the town where my four brothers and three sisters already work in the boot factory. I am Hungarian. But I have Russian teachers. They do not encourage the loser. So I will make boots and someone else will become the cellist.'

" 'Well,' I said, 'if that's all it is, don't let that stop you. I mean, I'll beat you anyway. Even if you don't kiss me.'

"He frowned. 'Why do you say this? I have heard that you play poorly this year.'

"I took a sip from his soda bottle. It was terrible, like medicine. I said, 'I'm not playing all *that* poorly. And you're not playing all that great. I doubt you'd win even if I were out of it. Your tone's very good in the middle register, but up high you have an edginess and your pitch is unstable.'

" 'Unstable? What do you mean?' he said, worried.

" 'You're sharp from high G on up. You just don't have command of that range up there.' I took another sip. 'And you know it, too.'

"He stared at me. 'Yes,' he said finally, 'you are correct in this.' He sighed and looked across the garden. Two men in dark suits were sitting with untouched glasses of hot tea at a table, looking at us. He waved at them. They did not respond. He looked back at me. 'I think perhaps I will kiss you after all. Perhaps tonight.'

" 'Great,' I said. 'Terrific.'

"That was that. He came to my room about midnight. And he kissed me. When he did I felt strong and cheerful, and also weak and stupid. I asked him to kiss me some more so I could see which side would win out. He kissed me some more. The cheerful side won. In fact, I was suddenly laughy and bouncy. That's unusual; you don't know me yet, but I can tell you I don't bounce a lot.

"We talked in between kissing. Milosz was clever and pretty generous. He was a country boy whose talent had popped up out of nowhere. When he was eight he was taken from his home to the city and stuck in with old teachers. They never treated his being young as anything but a sign for the need to hurry. I liked him. I didn't even mind that his pitch was off above high G. I was being clever and generous too. We kissed some more."

I shrug and flush—because of the memory, not because I'm talking about it. The talking is fine. "During the night I slept and woke up and went back to sleep at least twenty times. I didn't mind him breathing on me, or his arm under my neck, or anything. Usually I throw a fit if there's a wrinkle in my blanket.

"We woke up early because he had to sneak back into his room. I thanked him for spending the night with me. All he said was 'I am glad you won.'

" 'Won what?' I said. But he kissed me quickly and slipped out the door." I shrug. "Later that day I creamed him in the semifinal round."

My mother increases the touch from her arm on my shoulders, and orders a cognac. We don't say anything for a minute or two. When her drink comes, she takes a tiny sip and then looks at me.

"There's something else to the story, isn't there? I'd love to hear the ending if you want to tell it."

I nod, and take a sip of the glass of club soda the waiter brought with her brandy. "There were four of us in the round. I played third, and I played very well, finally. Milosz played last, just after me. We didn't get to see each other close up, but after I finished I sat down in the back of the hall to hear him play. Maybe if I'd stayed in the wings he wouldn't have done what he did. I don't know.

"Anyway, he started playing. I forget the piece— something pretty tame to warm up with. And he played it *terribly*, muffling easy double-stops, catching off strings in his draw strokes, really lousy intonation. It was frightening. He knew it was happening and at first he fought to get back in control, but things just got worse the harder he tried; when the piece was over he paused a long time before starting the next, and when he sat up again he had this peculiar tough look in his eyes. Someone else would have thought he had put his nerves behind him and found his resolve. But I guessed something different, and I was scared for him. I was sitting two rows behind his two guards.

"The next piece was a pretty tough scherzo from Dvořák, and he played it *gorgeously*. His pitch was almost perfect for a much wider range than before, and his edginess up high was a lot better. I sighed with relief. Milosz was okay. He played his next piece, the final one, just as well. Good enough for an Honorable Mention. In the later rounds, when the pressure's really on, if you correct an initial screw up as dramatically as he did, the judges sometimes reward you very nicely. If you're obviously not going to be in the running for the top prize, they feel more tolerant. An Honorable Mention would keep him out of the boot factory, I was sure.

"Milosz stood up as if he were going to leave the stage. But he didn't walk away. He was smiling. He looked my way. I went cold, then hot. My skin seemed to dry out and my dress felt like wood.

" '*Messieurs et mesdames,*' he said. 'I would like to play another piece for you this afternoon.' There was a whispering stir in the crowd. One of the judges stood and started to speak to Milosz, probably trying to save him, which was nice because they aren't allowed to speak to you at all, but Milosz waved him down. 'A brief selection,' he said. Then he giggled.

I knew he was a goner. 'From my heart, for someone else's future, with love.'

"Then he sat down and played—perfectly, too, wouldn't you know it?—'The Star-Spangled Banner.'"

My mother says, "Oh" softly, with a long release of breath.

"It was totally quiet when he finished. He sat there, and I think it really started to hit him what he'd done. He didn't look at me, but he pushed his fright aside, stood up, bowed to the crowd, and left with his cello. One of the cement-head guards was already on his feet. The other one turned, very slowly, and looked at me. Right smack in the face. His eyes were like little cups of bad coffee—murky, you couldn't tell what was under the surface, but you knew it didn't taste good. He held me like that for a couple of seconds—I felt like he had his hand very roughly under my chin. I couldn't move.

"Milosz didn't show up at the final round the next day. He didn't show for the awards ceremonies, either. They had given him an Honorable Mention. But of course it was too late now for an Honorable Mention to save him. Winning an H.M. for nailing the U.S. national anthem doesn't get you far in Hungary, I bet. It'll get you to the boot factory, and not a step further."

I shrug, drink some more soda without looking up, and crunch an ice cube. When I look up at my mom she's smiling slightly at me, but not with any big emotion. She's not especially charmed by my story, or amused, or proud either. Her expression is easy, friendly, cool—it says, *Interesting story. Yep, that's what we have to go through sometimes.* Then she looks at the waiter across the room and signals for the check.

23

I'm halfway through my second bran muffin with
bitter lime marmalade the next morning, and my third
cup of cappuccino. My mother is reading the ads in
a newspaper. She was reading the ads in another newspaper when I came down, and after getting up to make
my second coffee, she moved to this one, without
seeming to finish the other. I see a third paper still
folded, on a chair; she'll move on to that one in a
minute, obviously.

"Mom."

"Mmm," she says.

"Do you still like Indians?"

"Indians?" She looks up innocently. "You mean
like Sharash? Sharash did a nice city government
building for me eight years ago, but I hear he's gone
indigenous, building thatched-roof banks in Calcutta
and things like that."

"I mean American Indians. And not architects.

Weren't you . . . didn't you use to . . . study Indian culture?"

She looks truly confused; then a dark light flips on behind her eyes. "Did your father tell you that?"

"I asked him what you were like back then. What you were interested in. He said you were a student of Indian lore and history."

She snorts. "Oh . . ." She twitches the paper in her hands impatiently and makes a face. "I was a *hippie*. Indians were a hippie *thing*. That's all. It's not worth as much time as we're already spending discussing it—not that I mind, of course, I welcome your questions—but it was nothing. There were a lot of world-shaking fancies back then. They all passed, thank God."

"What about Melanie?"

"Who?" She has no idea.

"Macramé. What about macramé?"

She looks up again, trying not to seem impatient. "Macramé? God, your father . . . A good memory *is* revenge, isn't it?" She takes a last quick look at the ad she was reading, then looks back at me. "Macramé. Rope and knots. I did it for a while. I used to smoke pot and tie things. I barely remember. Why does it interest you?"

"You weren't . . . at the time I was born, you weren't hoping to expand your . . . craft into kind of a professional thing?"

She blinks. "I . . . well, who knows, I may have said something like that." She thinks for a minute,

227

watching me, and puts her paper down. "What are you looking for? You sound like you've seen a CBS special or something. Or had a nostalgic briefing from a former nonhippie who wishes he had been in on the fun."

"My father wasn't a hippie?"

She laughs. "He had too many scruples. And principles. And all the other words that end in -ples and mean you can't get loose and have a good time. Like *vegetaples*." She laughs again. "His idea of having a ball in the Sixties was giving up meat."

"He's still a vegetarian."

"I'm not surprised." She picks up the third paper and opens it.

"What kind of fun would you have that he wouldn't?"

"Any you can think of." She turns a page.

"Dancing in public places? Running around in the rain? Making noise that meant nothing? I'm just trying to get an idea."

"Sure. Yes. All that sort of thing. It's not like this was anything special to the Sixties, you know—every generation does the same thing. You and your friends do it too—you try things, find out what works for a good time, and then you grow out of it, or it leads to something nasty, and you forget it. Only your father would think it was something special to *his* times." She smiles and shakes her head. "Everything that happened to Cabot Spooner happened uniquely and for the first time." She turns to another section of the paper.

I eat a couple of bites. She reads a couple of ads. I ask: "Were you very clever to start your business?"

She looks up with a quizzical smile and stares for a second. "Yes," she says. "Damned clever." Back to the paper.

"How long after we left did you get started?"

She thinks without looking up, and says: "I started real-estate classes in the summer. About six months after."

"What were you doing in between?"

"Not macramé." She flaps a page over, and looks up. "I was exploring my options. I know that's a pat phrase, even a cold one. But I was facing reality, and reality offers a lot of paths." She pauses. "There were some unpleasant possibilities. Some things you and your father would never dream of."

She stares me down. I finish my breakfast, and she finishes her newspapers.

Later we go shopping for fancy groceries. My mother wants me to pick out my favorite foods. She is having a few people over to meet me tonight, and asks by the way if I would like her to invite some of the "Phrygian people." I tell her absolutely not.

We spend four hours getting an amazing array of tiny quantities of weird food. Especially herbs: We drive to three different stores for fresh herbs alone.

In between, we see a lot of the city and its surroundings. My favorite is Golden Gate Park, full of little nooks and statues. My least favorite is Sausalito, full of artists' studios that have been turned into stores selling brass ducks and $125 teddy bears.

My mother takes her job of showing me around seriously. I usually wear people out fast when I visit a new place, because I ask a lot of questions about details. But I don't wear my mom out a bit: She's ready for every question, and answers elaborately. Details matter to her, too. It occurs to me that maybe this is what she and Taxi had in common—an understanding of detail.

At one point my mother drives me over to Berkeley, but she doesn't show me the place the way she does other areas; she's getting somewhere, and the travelogue stops. I sit and watch, waiting. We drive through some choppy streets with old wooden houses, many of them being kept up just this side of a shambles. We wind around, make a lot of turns, get deeper into places that don't look good. Finally we pull onto a flat, short street full of tall, dark houses casting shadows on each other. Several of the lawns have sprung sofas sitting in bad grass. I see a few broken windows; after I glimpse a face disappearing from one of them, I stop looking.

My mother is still holding the wheel, staring straight ahead, clamping her jaw. She takes a big breath, relaxes her hands, and turns to me.

"This is where *my* Sixties ended," she says.

"You *lived* here?" I shiver. "It's not exactly my idea of Fun City. Has it gotten much worse since you were here?"

"It's gotten better." She points at a gray house at the end of the street. "That's my old place." There's a lot of broken glass in the front yard, and beside it a steep weedy hill with a rusty car stuck halfway up, like it was shot in the back while escaping.

We sit for a while. There's no sound, but I feel like we're being watched.

My mother points to the house across the street from hers. "A drug dealer lived there. Nice guy, very handsome. He started three free schools in poor parts of town, with paid lunches. He funded half a dozen small magazines for local writers and artists. He sold lots of speed. One night there was a quick shoot-out, and four people died. The police never even showed." She points two houses up from hers. "A women's group. They bought some of my pottery. Two of them published novels one year. All of them were busted; they were giving abortions. Some patients died. They went to jail for a long time. The yellow house to your left was the home of a famous comic-book artist who used to have block parties with famous rock bands he did album covers for. He would string the street with white paper lanterns on which he had drawn caricatures of all the people who lived here. I took a few art lessons from him—free, of

231

course. We went out a few times; he liked Walt Disney films. One of his political drawings was picked up by a big newspaper, and he started getting hassled by some authorities. He became convinced the FBI was bugging everything, shooting his house full of X rays from passing cars, poisoning his cats. One day he went nuts. Two of my housemates took him to a hospital and he never came back."

"Okay," I say.

"Just giving you the picture," she says. "This was where I started to weigh my options. The Sixties weren't turning into the Seventies with very much promise. I had just given up a big responsibility—you—so that I could take on another big responsibility—myself. I didn't want to waste my chance; it would be worse than criminal to squander the freedom I had insisted on." She motions toward the street. "This was one option—the so-called counterculture. I was doing a lot of pottery, selling a few pieces. One of the people here wanted to start a cooperative workshop; he offered me a job. Someone else was manufacturing incense and I could invest in that, too. But just look. Feel this place. People were *dying* here—and they were the *successful* ones."

I shiver again, and glance at the car on the hill. "So you got out, and went to real-estate school?"

"I didn't get out at first. I tried to mix both worlds. I still thought I was a hippie, smoking dope and sleeping with anyone in my house. Things that *used* to be fun. But it was obvious: in one part of my world people were moving ahead, and in the other part they were moving less—in *any* direction—every day."

She starts the car, and manages to turn around in the street, avoiding the driveways. We head back toward San Francisco without saying much. When we arrive home, I peer up and down the street for Martin's BMW. It isn't there. When we get inside, my mother puts her hands on my shoulders at the foot of the stairs and says, "I'm so sorry—I've run you ragged. You seem so fresh that I keep forgetting you just got jostled across the country, probably without much sleep on the *ground* every night."

"I slept fine on the ground," I try to say, but a yawn makes it incomprehensible. She nods, and says, "Why not take a bath and a nap? I'll call you in plenty of time to dress."

I stumble upstairs and shuck my clothes onto the floor. Just as I'm yanking back the sheets on my bed, I notice something I would hate to have missed: there's a blue cornflower on my pillow.

24

"My son," says a lady in a blouse with a bow out of a Kodak Christmas-morning ad, "has the most beautiful ear."

The philosophy professor from Stanford giggles and pulls on his lobes. The woman goes on. "He picked Kreisler songs out at fourteen on the violin. We bought him a marvelous instrument. Not a Stradivarius, but an Italian, definitely Italian. We *leapt* up the waiting list at a very renowned violinist's studio. My son began studying with the man, and was soon picking out *Schubert* tunes." I almost tell her that Schubert tunes are in fact easier to manage than Kreisler, but instead I just listen to the inevitable: the boy had such promise, but seemed all of a sudden to let his mind wander from music. He stopped practicing very much. Soon he was cutting lessons.

"Next," says the philosophy professor, who so far in one hour of intruding on my dozen conversations has not said anything nice *or* funny, "will come to-bacco, and pool halls, and buxom women, and the dialectic. After that . . ." He clucks his tongue.

I hand him my glass. "Get me a drink, will you, Plato?" As he grumbles off I tell the woman that her son has reached the hardest time for a young musician: the year between fifteen and sixteen when everything in his world and his body is clamoring for his attention. I tell her to ease up, not to press him. To even suggest that he drop back to two lessons a month. "The most important thing is not to let the music come between you at a time when alienation is *so* precariously easy," I say. She thanks me devoutly, pressing my hand. The philosophy prof returns at that moment with my ginger ale and wails, "My God! Careful of her hands! She's a genius!"

I sigh, and turn dutifully toward the next guest in line to chat with me. It's Martin. He smiles, hands me a glass of wine, and says, "There's someone over here I want you to meet." I hand my ginger ale back to the professor and let Martin guide me through the room, down a hallway, and around a dark corner into an empty room.

I put my hands on his chest. He's wearing a nubby tan silk sports jacket with the collar up and the sleeves rolled, a pale-purple shirt, a darker-purple silk tie, and black pleated pants, very baggy. He looks nineteen tonight, and so handsome my head aches.

"I have to be careful," he says.

235

I freeze. "Why?"

"I don't want to drive your music between us. Alienation is so precariously easy at this tender time."

"That was last year." I laugh. "Between fifteen and sixteen, remember? I got all of that stuff out of a psychology journal article one of my concerned fans sent me in the mail."

He puts one hand over both of mine on his chest. "What happens between sixteen and seventeen? For the well-adjusted child musician?"

"Love. And I'm already behind."

He laughs, but also bends his head forward and gives me a kiss—a tiny one, only a proper peck really, but enough to make me gasp and try to free my hands to pull his head down for another. But he holds my hands on his chest, and stands a bit more upright.

I groan. "How much longer do I have to talk to these people? Can't we get away from here for a drive or something?"

He nods. "Stick around for another hour or so. You're wonderful at this. Your mother is very proud, and her friends are good people. They like you."

"If I know we can take off, I'll do it." I sigh. "You'd better release my hands now—I'm a genius, you know." He lets go, and I touch his cheek. It's dry and smooth.

When I get back to the party I feel like I'm shining all over. I kiss my mother on the cheek, pump the hand of the architect she is only halfway through introducing me to, and grab a handful of pesto canapés and prawn toasts and Brie mini-soufflés. I tell the architect that he is right, sometimes Brahms is more moving to play than Beethoven. Fifteen minutes later I tell a woman who paints art deco murals that, indeed, she is correct—Beethoven is often more deeply felt than Brahms. I assure a stereo designer with two pieces in the Museum of Modern Art that I see his point about the sound quality of a system being related to its beauty, and I applaud the insight of a bamboo-thin woman poet of mixed black and Oriental lineage when she tells me she's certain I often think of random words when I'm playing instrumental music. I grin, I laugh, I touch people on their arms, I moue, I meow, I damn near turn flips. I am a charming, witty, star of a daughter. My mother whirls me around from spot to spot, even retracing steps with people who had used up their ration of me before my shift into gaiety. She squeezes my hand. She tells me I'm doing great. She feeds me Frangelico truffles and dried baby plums with a Grand Marnier glaze. On my own I grab a couple of flutes of champagne.

I get so dazzled with my own performance that when I feel a polite pluck at my sleeve during a talk with a homosexual diamondsmith wearing a tie clip that says FUCK AIDS in twinkling chips, I don't even know it's Martin behind me.

He says, "Excuse me. I've been wanting to ask you all evening. Do you *really* think Mozart wrote all of that music when he was seven?"

237

The diamondsmith slips away with a groan, and miraculously, no one else is there to snag me. I turn to Martin. "No—he wrote some of it when he was six. Let's get out of here." We zip through the front door and angle across the lawn away from the big windows in the living room. "I parked up this way just so we could cut out quietly," he says.

I take his arm. "Is this terrible of us?"

"I'll cool the boss."

"What are we going to do?"

We get to his car, and he swings open my door. I smell his sandalwood cologne as I slip by him. He goes around and climbs in his side, and says, "You served your time with the parent class. Now we get to be youth."

He starts the engine, lets it warm for a minute, then backs up and pulls away. A tape plays Ravel piano music softly, *Le Tombeau de Couperin*. He reaches to turn it off but I stop his hand, and hold it for a moment, until he needs to shift.

Our drive takes us deep into the city, deep into Friday night. Everywhere in the streets, people are *playing*: I can see it in the way they open car doors and smile, looking up or down the block. We go past fancy Cajun restaurants and funky Tunisian restaurants and theaters running Japanese film festivals and bars featuring two-for-one St. Pauli Girl specials announced on sandwich boards hung over scarecrows propped on the sidewalk. We go past bookstores with tiny open doors showing golden-brownish light; inside I glimpse people standing still with their hands on their hips or arms crossed, looking intently at bookshelves above eye level—and I can see in a flash that they are playing too, in their way: *It's Friday night, and I can read any one of these books.*

"What is it about Friday night here?"

"A week of work is over. But once Saturday morning starts, the clock is ticking on the weekend—it will soon be over, and Monday looms. Friday night is like an extra gift. A free space. A stolen interlude."

I nod. He looks over. "What do you usually do on Friday nights?"

I have to think for a moment. "I have an improvisatory music group that plays Friday night. Every week." I look around. An Ethiopian restaurant and a double feature of grade-B gangster flicks zip by. "Every week I'm in a basement in a Silver Spring ranch-style home where a twenty-seven-year-old pianist still lives with his parents, playing weird abstract music that mostly goes nowhere." I laugh incredulously.

"Do you think that's perhaps best? That you might waste your time otherwise, on gangster movies and imported beer?"

"I think it's Friday night." I smile at him. "And you look like just the guy to steal an interlude with."

He smiles. We drive on for another ten minutes in silence, and finally pull up on a quiet street with nice old and new buildings. Signs hang on poles and over doors, steps lead up or down to doorways giving out dim colored light—some bluish from below, some orange from above, some flashing gently in neon colors. "Bars?" I ask.

"They're called 'clubs,' but . . ."

"Sure. The difference between a 'club' and a 'bar' is the difference between a 'prawn' and a 'shrimp.' "

He laughs. "You're already hip to this city."

The easy way he says it makes me look at him. He's glancing up the street to pick his spot, with quick, eager eyes—a boy on a date. Which makes me a girl on a date. I laugh.

"How does one choose?" I ask.

He glances at me. "By the kind of music, usually. Not the type you play, though. Mostly it's rock, different kinds of rock. What do you feel like?"

"I have no idea. I don't know any kind of rock."

His eyebrows go up. "No? Then that makes it easy. We'll go to Penny Lane."

"What kind of music is at Penny Lane?" I ask as we start walking past the doorways, each of which puffs out a neat blast of loud stuff.

"All kinds," he says. "They have a jukebox, not a band. It's quieter. The kids are more into talking instead of drinking and dancing. That sound okay?"

"Yes. Perfect."

And it is. We spend two perfect hours there. The place itself is really nothing special: a nice room with rough stone walls, old wooden booths, new wooden tables, exposed beams in a stone ceiling, and posters of old rock groups. What we do for those two hours is nothing special, either: we talk. Not so much to each other—we talk to a lot of kids, who shuffle in and over to our table and out again, all of them knowing Martin, all of them greeting me nicely, all of them fresh and interesting and loose.

I sit and listen, and watch these kids—they're all around eighteen or twenty, I guess—and they don't care that they're not "accomplishing" anything tonight. I'm *always* "accomplishing" something. And usually I hate being with people who are wasting time. But here I am, with these friendly, smart kids. They aren't sucking down the booze, or chattering with fake drama and giggles, or chasing each other. They're just playing calmly on a Friday night. And I'm playing with them. Yes: I'm not trying to squeeze a counterpoint to Effie Peeters's scrawny oboe improvisation in between flat Roger on trumpet and sharp Alan on violin; I'm not studying a score. I'm having *fun*.

There are all kinds of kids here: punks, slicks, cleancuts. Some wear sleeveless leather shirts and have dyed stripes in their hair, some wear oxford-cloth buttondowns and tight ties, some wear sweatshirts that say U.C. SANTA CRUZ or tee-shirts that say TALKING HEADS REMAIN IN LIGHT TOUR, SUMMER 1981. I gather a lot of them go to art schools in the city.

I find myself tapping the table a few times to certain

songs while we talk. At a lull in one of our talks—about some new architecture in Marin County—a boy with black spiky hair, a piece of dental floss tied through a hole in his ear, and gloves with no fingers asks me very politely if I'd like to dance.

"I don't think I've danced since I could walk," I say.

"Cool," he says. "*Au naturel*, then." He grins and takes my hand and leads me over to the small dance floor.

"I don't want to do the standing-alone-gyrating kind," I say. Without a hitch he twirls me by the hand into a spin, then pulls me into a loose clutch and leads me through easy, rhythmic steps. He's smooth, and he laughs when we do something well or poorly. When the song's over he holds my hand and walks me back to the table, thanking me for the dance.

During the drive home, Martin asks: "So how was your first Friday night?"

I look at him. "I'm hoping it's not over," I say. He smiles slightly, looking at the road. I lean back and stretch my legs. "So far, it's wonderful. I can't believe I danced without having any idea of what I looked like."

"Are you that insecure about your looks?"

"No. But I'm used to rehearsing before performing."

"Dancing at Penny Lane isn't exactly performing."

"No. And I guess *that's* what's different. I never do

242

anything that isn't performing or preparing to perform."

"That's hard for me to believe. You're the least self-conscious person I've ever met."

"Really?" I laugh and blush. "I never thought of myself as being unself-conscious."

He laughs. "Exactly."

I think about it for a minute, watching him drive. "I know why you think I'm so spontaneous. It's because I feel happy." He says nothing. "Want to know why I feel happy?"

He glides onto a straight, dark road that takes us high above the twinkling bay. "I think I know already," he says.

"I bet you do." I close my eyes to say it: "I'm in love."

He smiles, glances at me, and says, "I'm glad you realize it."

This sure disappoints me. "That's a weird response," I snap.

He touches my knee and smiles. "I'm sorry. I suppose it's a cowardly way of saying I think you're only half right."

"That's not too brave either. You want to tell me what you mean?"

He nods, unflustered. "You're right in thinking you're happy because you're in love. But you're wrong in thinking that it's just happened in the past two days, here."

I feel like he's speaking another language. I can't come up with a good question, but he does.

"Where do you think you get your feeling that you can do anything? Your spontaneity, your talent, your freedom from doubt?"

"That's easy enough. I grew it. Myself."

"Ah."

"You don't think so?"

He smiles over at me in the dark. "Well, no, I don't," he says. "I think maybe it's a mistake to take your self-sufficiency that far. You deserve what you have, yes. But you *have* it because of love. Loving someone is the only way you get that kind of energy and daring. That kind of freedom. And you've had that for a long time, Sib."

We pull onto our street and stop. He turns toward me, puts his hand gently on the back of my neck. "I'd probably have a wonderful time letting you think you were falling in love for the first time, with me. We may have a wonderful time anyway. But first—you should face your best-kept secret."

With that, he pulls me over and hugs me against his chest, and I start to cry, very quietly. With a strange, sudden feeling of relief, I cry for a long time, more and more softly, and he lets me, until finally, I fall asleep. I wake up once, carried in his arms like a four-year-old, as he unlocks the door to the house, but I am asleep again before he takes me upstairs and tucks me in.

25

I jump out of bed at five in the morning and land flat on my feet and realize suddenly: I have an audition and I haven't touched a cello for almost three days.

I wash and put on jeans and a shirt and sneakers and I get out the Bianchi. I play, loudly, and it's not exactly music. I've been at it for fifteen minutes when there's a tap at my door. I don't answer. Another tap, then a knock. My mother says, through the door: "Sibilance?"

"Good morning!"

"Do you have to do that now?"

"Yes."

"Really? It's 5 A.M."

"I have to warm up. I should have been playing some of the time I was hunting for silk sashes and fresh coriander. I'm cold, my cello is cold, and the audition is in seven hours."

I hear her yawn. "All right. But—it's *awful*, what

you're playing. Can you possibly make it quieter?"

"I'm afraid not."

She yawns again, longer, with a little sigh at the end. "Well. I guess it's good morning, then. I'll go down in a while, and we can have breakfast when you take a break." She goes.

But I don't take a break. I warm up the instrument, then I practice; then I play; and then I noodle. I touch on nearly every solo piece I know, gently, not putting too much pressure on my fingers, but getting rid of the softness that had already started to creep in. I lose track of the time.

My mother knocks on the door again at 11:15. "Sib, are you dressed for the audition?"

"No."

"Perhaps you should think about it?"

"Okay."

I don't think about it much. Out of habit I put on my underwear and white stockings, and pull my black dress out of the closet. But as I take it off the hanger, I do think—and put it back.

When I come downstairs, my mother claps her hands softly and says, "Oh, I'm *so* glad you changed your mind." She smiles and hugs me lightly. "I do think you're ready for red silk. From now on. Let's declare the days of black officially over."

During the drive she points out a few things. I don't say much, mostly grunting as politely as I can; I don't like to talk before I play. She gets the idea after a while, and we ride for a long time in silence.

When we're almost there, she says, "Sib, you know that I want you to stay with me."

"Do you?"

"Of course I do." She looks at me. "I'm worried that . . . well, I *could* have lied to you the other night, you know, told you my decision to let you go, back then, was based on something less . . . selfish, but . . ."

"My decision won't be based on your decision then." I touch her on the arm. She tries to smile. "I like you, Mom. I don't hold any grudges. You make good sense."

"Are you sure? You're not deciding anything because of the past?"

"I'm sure."

"I suppose," she says, as we turn into a green glade with a road winding down, "that you'll base things largely on the audition. Is that right?"

I hesitate. "That's where I'll decide," I say.

She gives a little sigh. "I've been a bit premature, I guess, in thinking that your coming all the way out here—such a major move—represented a decision already." She means it to be a question that doesn't look like a question. So I give her an answer that doesn't look like one either: silence.

The Institute suddenly rises out of a small, grassy valley full of what look like dogwoods and fir trees.

247

There are a few modern white buildings arranged obliquely. I would guess one is for classrooms, one for practice rooms, and one for recital and concert halls.

"Nice place," I say, as we get out of the car.

She is gazing fondly at the valley and the buildings. "Isn't it beautiful?"

"Did you broker this, or whatever?"

She nods.

"Did I meet the architect last night?"

"Yes," she says. "It was Martin." She smiles at me, puzzled. "Didn't one of us tell you?"

"No."

She smiles, and looks back at the buildings. "He won an open competition with these plans, when he was only nineteen. But this is it—he's done nothing else." She sighs. "People thought he would really take off. But he says he doesn't have what it takes, yet." She watches the buildings, as if they hold a clue. "It's been a few years. I hope he gets it soon."

"Let's go in," I say. We go down a stone path leading to the first building and walk through large redwood doors. Inside, a man in a dark-blue suit greets us, shakes my mother's hand, pulls his goatee, and bows to me. I curtsey, and he blushes. My mother introduces us: he's the dean who wrote me the letter.

My mother says she'll go take her seat. She gives me a kiss on the forehead and an awkward squeeze, and goes.

"Miss Spooner . . ."

"Sorry about the curtsey," I say. "I have a sarcastic streak when I'm nervous."

"Not at all, I thought it perfectly charming," he says, but I can see he's relieved. He has a Virginia accent. "I'm delighted to hear you're a little nervous," he says, with a smile. "It means that our audition holds for you perhaps a small part of the importance it holds for us."

"I like to perform well."

"And you do, you do. As everyone in our world knows. I'm certain you will astonish us today."

"I'd like very much to astonish you."

"Delightful," he says. Then he asks me if I would prefer to meet the attending faculty now—they "await" me in the greenroom—or after the audition. I tell him I'd prefer to meet them later.

"Of course. Then after I show you to your rehearsal space, I'll excuse myself to tell them your preference. Then I'll return to notify you of the time ten minutes before the audition is to start, and again at two minutes before." I thank him. He takes me to a little room with a music stand, and leaves.

The trick now is to think only of music for twenty minutes—not to think of the people here and what they're looking for. I feel a curiosity seeking me in the building like a sleek bat zeroing in on a moth; it's Dzyga, and Martin, and my mom, and even Taxi, who's seen me so many times. I have to dive into the middle of the Schumann "With Humor" piece to drag myself away from thinking about all of them now, maybe even from crying a little. A red dress is fine—but no red eyes.

I finish the Schumann and do some pizzicato exercises I pulled from Frank Martin's second Étude for string orchestra. My hands pluck and hop so enthusiastically that I decide to include a pizzicato piece in the recital, a part from the Debussy String Quartet. Part of the fun of playing your own program is using a particular skill that happens to be hot at the time.

A tap comes at the door. "Thank you," I call. Then I play a long windy run from Bridge's *Oration*. It's always the last thing I play before going onstage. Then I sit in silence until the next tap.

The dean, formal and proper, doesn't meet my eyes as he guides me through a short backstage walk to the wings. He leaves silently, and I look past the single chair onstage and up into the left wing of the auditorium's seats, my heart bumping. Three faculty members sit within my view in the third and fourth rows; beyond them, another ten rows back, on an aisle seat halfway between stage and rear wall, sit my mother and Martin.

Complete quiet falls on the auditorium: the dean has made some signal that I am coming.

A faculty does not applaud at a recital. The role of judge sobers everyone. At Juilliard I was surprised by a quick burst of applause as I took the stage, and almost unnerved; here the members are silent and watchful.

I walk to the chair, holding the Bianchi, and face them for a moment. There are twelve, all sitting in the third and fourth rows of this low, bowl-shaped auditorium. Four women, eight men, half young and half old; but Dzyga, I am certain, is not among them yet.

I take my seat, making no noise, even when I set the point of my cello. When you give a recital you cast a spell, a spell that must begin from the moment you walk onstage. You want to seem to your judges to be something pure and mysterious. After the first sign you give that you are merely human—a cleared throat or a scraped chair or a smile—they will warm to the task of finding other signs, and the easiest signs to notice are your mistakes. *Hold your spell and they will miss your first three mistakes,* Gustavus told me; *more than that you shouldn't make.*

The performer can choose to announce the pieces or not. I never do. Your voice is the worst spellbreaker there is. I let the silence, which I disturbed by my quiet entrance, rise fully around me and the Bianchi and reach to the back of the hall. Then I play.

I begin with the windy Bridge passage, full of sonority but pricked with odd quarter tones and sixteenth notes. More than anything I play, it establishes like an overture my tastes and strengths: it's difficult and bold, but still loyal to the traditions our ears love

251

beyond memory. The piece is oblique, but digs at feelings precisely; its subtle structure cheers the intellect. What a guy Bridge was, and what a gal I am: I *blaze*, and when the Bianchi roars out the last throbbing low note, I feel a collective gasp.

Next I play Haydn: my own solo version of the first and second movements of the *Kaiserquartett* in C, the greatest string quartet in history. Reverence and simplicity, but with zest. When I finish the cantabile on three leading notes, I feel regret from the listeners—they want the menuetto to follow. Good. Gustavus: *Let them want more of the same, and give them different.*

As I prepare to play the Debussy pizzicato piece, I suddenly feel ice in the air. I shouldn't look up—it can cut the magic—but the Bridge and Haydn have been so dazzling that I can risk it. I raise my eyes, and see a quiet entry: On the right, Dzyga has come.

He's in no hurry, though he's not rude: the three steps I see him take, the lowering and inspection of his seat, the sitting down, and the arrangement of trouser knees make no sound at all. Then he raises his eyes. And we stare.

He's tall—at least six two. I didn't expect that. His hair is white on the top and back of his head, above a high, straight forehead. I thought he'd be short and prancy, with the usual thick, wavy hair of the dark Russians. His complexion is dark, as are his straight, clipped mustache and his thick eyebrows. A surprise: he's got Tartar in his blood somewhere, with that tight skin around the eyes that gives the illusion of slant. It's exotic, secret, dangerous. At thirteen he must have scared the shorts off every prawn-pink cello boy who came to those competitions from Vienna and London and Cincinnati.

When he gives me the very slightest of nods as a signal that he is ready for me to resume, he almost scares *me*. But not quite. In fact, his air of authority in giving me the nod piques me, and I decide to throw a little right back. So before his head has really finished its tiny motion, I wham into a dissonant chord and leap out of it into a squealing, high melody played sul ponticello. Then I peel back into some bass theatrics that resolve into a long-toned rhapsody reminiscent of sweet Bartók. The piece isn't by Bartók, however. It's by Spooner: I'm making it up as I go. Not exactly a standard recital practice, but I'm feeling powerful today.

I wheel out of the rhapsody's river into a little forest of pizzicato—inevitably derived from the Debussy I almost played and the Martin I rehearsed, but still pretty fresh and unusual. From there I suddenly see my way out, with glissandi into a recapitulation of the rhapsody theme in shorter, higher tones, and then in double- and triple-stops. These carry the melody into a final chord as booming as the first, but euphonic, resolved. As an afterthought, I pluck three quarter notes that quote from the Haydn menuetto. Several of the faculty members laugh.

I've never been hotter. I know that I can end at any time now, whip through the Szymanowski as requested and leave them incredulous, but this is one of those days when every nuance of technique and intellect and emotion finds it way into the fingertips. I sail through a movement from the second Beethoven sonata as a kind of cool wind, then jounce through a transcription from Kurt Weill's violin concerto. Then a French medley: tunes I've picked and pieced from Honegger, Milhaud, and Chausson. A final heartbreak-beauty piece to precede the Szymanowski: Brahms's *Alto Rhapsody*, which cellists are fools to avoid.

I look up one more time before the Szymanowski. The faculty members in the front are completely dazzled—I can see everything from barely suppressed exultation to stunned incredulity in their faces. My mother is looking lost, as if she can't find her way to pride through her astonishment. Martin is smiling like he knew it all along. As for Dzyga—his eyes are closed. I'm amazed, almost insulted at first. But then I realize it's probably the rare response I never even hoped for: blind immersion in the music.

As I swing my eyes back to my task, I catch sight of another dark figure dead-center in the very back row. Taxi, sitting where he always sits at my recitals.

His being there is as comfortable for me as the neck of the Bianchi; I have not forgotten him today.

Zip, slash, whizz, boom—the Szymanowski erupts. I am everywhere and everything on the instrument, high and low, quick and slow, slippery and chopped. Taxi is right—it's an encore show. But it's great, and it wraps up the whole recital with passion and fizz and melancholy and holler. I close my eyes too, and instead of feeling myself *play* the music I seem too to be a listener, evoking the music by *hearing* it a millionth of a note before it comes. The last race of triple stops shrieks to a peak, and the crescendo falls. I snap the bow back across in a flourish of melodramatic chord—and the faculty explode onto their feet, pounding applause that makes them sound like hundreds, cheering and calling. I raise my eyes, and see things. My mother on her feet grinning and weeping. Martin rising with a bouquet of flowers in his hands. Dzyga sitting motionless in his chair, tears glinting wet on his cheeks in the faint light, as his eyes stare not at me but beyond the Bianchi to someplace we have both been. And Taxi rising from his seat and making his way unobtrusively toward the end of his aisle and an exit.

I look at these four people. I thought at first I wanted something different from each of them, but really I wanted the same thing. And I have got it. Each of them is giving it; it's absurd to choose among such complete offerings. But I have to choose, without fear of losing anything. At last I see that I can't base a choice on how I think they feel about me. The difference is in how I feel about one of them.

The applause pounds. Martin arrives at the edge of the stage and looks at me. I meet his eyes; they are bright and full of promise. He holds the flowers up. The applause renews. But I do not move to get the flowers. And I do not bow and leave the stage. I hold up my hand.

The applause stops. I sit back down, and set the Bianchi.

26

My mother catches Taxi as he walks back to the bus with a cup of coffee from a dispensing machine he found in the classroom building. Two minutes later he would have been gone.

"Cabot," she calls. He looks up and stops five feet from the bus, holding his paper cup and waiting.

"Hi, Connie," he says.

She's panting a little. She stares at Taxi, and he lets her catch her breath.

Finally she says, "Quite a finish, wouldn't you say?"

"The Szymanowski? Yes, it's pretty astonishing. So was everything today. I liked the improvised piece best. You must be knocked out, hearing her for the first time."

She watches him doubtfully. "I'm not talking about the Szym . . . about that. I'm talking about the last thing she played. That's what 'knocked me out,' as you say. And just about knocked Sib out, too."

He looks surprised. "She played an encore?"

"Are you telling me you weren't there?"

"Yes," he says. "I've been taking a look around the campus here. I never stay for her encores. She asked me not to. She says . . ."

"This one you should have stayed for." She eyes him carefully. "Are you trying to make a fool of me, Cabot?"

Taxi seems alarmed and also almost angry. "Tell me what you mean, Connie."

My mom sighs heavily. She doesn't seem to be able to decide between anger, hurt, and confusion. She asks: "Where were you going to go, just now?"

"Home."

"To Washington?"

"That's where I live."

"Alone?"

He hesitates. "Yes."

"Aren't you sure?"

"I'm sure," he says. "I just haven't gotten used to the idea."

"You were leaving her here with me."

"I assumed you two had arranged that." He looks at her sharply. "Don't gloat, Connie."

"You misunderstand. We don't gloat in the boot factory. I don't suppose that means anything to you— the boot factory?"

He shakes his head and studies her, frowning. "I'd

better go, Connie," he says, and takes a step toward the bus.

"Wait." My mother is rubbing her eyes with one hand. She looks up, back at the school, then faces Taxi. "I'm not being very clear. I find it hard to believe that I have to explain this to you. I find it hard to believe I wasn't set up, all along. But I'll try."

Taxi says nothing.

"All right." She takes a deep breath. "If you left when everyone was applauding, you missed something. Something I hope you can interpret for me, if you'll do me the favor." She looks back at the building. "I don't suppose you know where Sibilance *is*, do you?"

"Probably in a rehearsal room somewhere. She goes off by herself to come down to earth."

My mother nods, looking around the buildings. She finally turns back to Taxi. "After you left, then, Sib made a signal for the applause to end. Then she said she was going to play an encore." She stops and studies him. "Couldn't guess what it was, by any chance?"

"No. I told you, I never . . ."

"Yes, you told me. All right. I just thought you might know this 'piece.' Because, after all, you *wrote* it."

It's Taxi's turn to look confused. "Wrote it?"

" 'I'd like to play a final selection,' she said. 'It's an

original composition by Cabot Spooner and Sibilance T. Spooner. It's called *The Peace and Love Shuffle*.' God," my mother says disgustedly, "you had to grind that hippie bullshit into her, didn't you?"

Taxi is flabbergasted. "She played *that*? In *there*?"

"So you *do* know it."

"Well, sort of. I mean, it's kind of an exercise we did in the bus on the way out. But . . ."

"When she was finished with it, she did what I believe is called a 'segue' into a song I happened to recognize because it was one of your anthems in those grand old days—'In the Midnight Hour'?"

Taxi is amazed, shaking his head, smiling incredulously. My mother watches, nodding. "I thought so," she says. "Oh, God, what am I going to do with her now?"

"Connie." Taxi holds up a hand. "Listen to me. I'm as surprised by this as you are. Nobody's making a fool of anybody. I think . . ." He takes a deep breath. "I think Sib was just saying goodbye."

"To you?"

"Of course to me."

"After you were already gone?"

He shrugs. "Sib does things for principle, Connie. This was a goodbye for principle. And"—he smiles— "as we see, the message *did* get through."

"You take it only as a goodbye, then? And you're happy with it? Not going to change your mind?"

"I'm not 'happy' having Sib say goodbye to me at all. If you need to have it spelled out for you, I'm probably as unhappy as I could ever be." They look at each other for a few seconds. Then Taxi says, "But this helps a little, as I'm sure Sib intended it to. Your time of being alone is over, Connie, but mine is just beginning, and this is Sib's way of easing me into it."

"And easing herself out of the Phrygian Institute."

Taxi gapes. "What do you mean?"

She laughs coldly. "Well, can you blame them for being upset? They assemble out of session for this precious prodigy who deigns to give them an audition, and they respond to her quirky recital with incredible warmth, and she repays their welcome with a bizarre hippie improvisation and a trashy AM radio tune."

"But . . . they turned her down?"

She watches him through a deep breath. "No."

"Good." He nods. "That would have been very foolish."

"I'll tell them you said so. No, they just *almost* dropped her. In a rage. Most of them were firm—the insult wasn't lost on them. This is a serious academy, not a place for audio valentines. But I talked to them, and, most importantly, the new cello professor talked to them. Very forcefully. He's a big shot from Russia; they pulled a big coup when they got him to defect, I gather. He was adamant about Sib, thank goodness:

'If she is denied, I return,' he said. No one could talk him out of it. They ended up believing him, and they gave in."

"So she's in."

"She's in."

"Does she know? That's she's in?"

My mother hesitates. "I don't know. We . . . no one could find her after she left the stage. My secretary went to the rehearsal room she used before the recital. Her cello case was there but she wasn't. He's checking the other rooms now."

"Well, then. I'll be going." He opens the bus door, still holding his coffee.

"Cabot."

"Yes, Connie?" He's settling into his seat, balancing the coffee, putting the key into the ignition.

She looks up at him, open and pained. "What am I going to give this girl? I mean, from *me*."

"Whatever she wants, Connie. She's fair, and she'll take it."

"If she can find it."

"Oh, she's found it already, Connie, or she wouldn't be staying. Sib doesn't take chances like that."

He starts the bus. My mother, looking convinced, watches as he warms it up. He sticks his head out the window.

"Will you do me the favor I did you?" he asks.

"What's that?" she says, looking lost.

"If she asks to see me—once—will you send her?"

She looks at him hard. "If she asks. Of course."

"Goodbye, Connie. You're a very lucky woman." He puts the bus in gear.

"I'll need the luck. God," she says, looking suddenly desperate, "I don't know if I can *do* this, Cabot!"

Taxi backs out, and putts off up the road through the glade, and around the bend. My mother watches until he's gone.

And fifteen minutes later, as he edges into the express lane of the highway heading east and settles the bus into its low whine of fourth gear, he hears, above the clean little engine and the zing of the tires on the road, the sudden loud groan of the Bianchi, as it fills the bus with the first of the four strung-out chords that start that midnight masterpiece. By the third chord he has pulled across two lanes to a screeching stop on the shoulder, and then, just as the fourth chord has droned out and the melody is about to begin, he scrambles into the back, and takes that girl, and holds her.

Bruce Brooks grew up on the East Coast and graduated from the University of North Carolina and the Iowa Writers' Workshop. He is the recipient of the 1985 Boston Globe–Horn Book Award for Fiction for his first novel, THE MOVES MAKE THE MAN, which was also a 1985 Newbery Honor Book. He lives with his wife and son in Silver Spring, Maryland.